D1072574

QUANTUM PHYSICS
FOR
POETS

QUANTUM PHYSICS
FOR
POETS

LEON M. LEDERMAN
NOBEL LAUREATE
CHRISTOPHER T. HILL

59 John Glenn Drive
Amherst, New York 14228-2119

Published 2011 by Prometheus Books

Inquiries should be addressed to
Prometheus Books
59 John Glenn Drive
Amherst, New York 14228–2119
VOICE: 716–691–0133
FAX: 716–691–0137
WWW.PROMETHEUSBOOKS.COM

15 14 13 12 11   5 4 3 2 1

Library of Congress Cataloging-in-Publication Data

Lederman, Leon M.
Quantum physics for poets / by Leon M. Lederman and Christopher T. Hill.
p. cm.
Includes bibliographical references and index.
ISBN 978–1–61614–233–9 (cloth : alk. paper)
1. Quantum theory—Popular works. I. Hill, Christopher T., 1951– II. Title.

QC174.12.L4326 2011
530.14'3—dc22

2010044163

Printed in the United States of America on acid free paper.

*Leon dedicates this book to his wife,
and super-assistant, Ellen.*

*Christopher dedicates this book to
Katherine and Graham.*

CONTENTS

ACKNOWLEDGMENTS 11

CHAPTER 1. If You're Not Shocked, You Haven't Understood 13
An Advance Peek at Quantum Physics 16
Why Is Quantum Theory Psychologically Disturbing? 20
Spooky Actions at a Distance 28
Schrödinger's Tabby 31
No Math, but Perhaps a Few Numbers 33
Why Do We Care about a "Theory"? 35
Intuition? Fire Up Your Counterintuition 36

CHAPTER 2. Before the Quantum 41
Complicating Factors 42
The Parabola and the Pendulum 46
The Cannon and the Cosmos 48

CHAPTER 3. Light and Its Various Curiosities 55
How Fast Does Light Travel? 57
But What Is Light Made of? Particles or Waves? 58
Thomas Young 61
Diffraction 63
The Ecstasy and the Agony of Young's Double-Slit Diffraction Experiment 66
Young's Conclusion: Light Is a Wave 71
Unanswered Questions 72
Fingerprints of the Atom 74
Maxwell and Faraday: The Laird and the Bookbinder 76

CHAPTER 4: Rebels Storm the Office 83
What Is a Blackbody and Why Should We Care? 84
Ich Bin Ein Berliner 88
Catastrophe! (in Ultraviolet) 90
Max Planck 91
Enter Einstein 94
The Photoelectric Effect 96
Arthur Compton 101
The Double-Slit Experiment Returns with a Vengeance 102
Booby-Trapping the Slits 108
Through a Glass, Brightly 112
The Walrus and the Plum Pudding 113
The Melancholy Dane 115
The Character of the Atom 117

CHAPTER 5: Heisenberg's Uncertainty 119
Nature Is Lumpy 119
The Franck-Hertz Experiment 123
The Terrible Twenties 126
A Strange Mathematics 129
The Inception of the Uncertainty Principle 132
The Loveliest Equation Ever Written 133
Fourier Soup (or, I Think We're Back in Kansas) 136
Waves of Probability 139
The Triumph of Uncertainty 142
Born, Fourier, and Schrödinger 143
The Copenhagen Interpretation 144
Still Crazy After All These Years 147

CHAPTER 6: Quantum Science at Work 149
But Isaac Newton Never Sent Us E-mail! 150
Playing Cards with Dmitry Mendeleyev 152
Police Lineup of the Elements 154
How to Build an Atom? 163
The Atomic Orbitals 165
Enter Mr. Pauli, Stage Left 169

Molecules 173
Summing It All Up 176
Pauli's New Force 178

CHAPTER 7 Controversy: Einstein vs. Bohr . . . and Bell 181
Four Shocking Things 183
How Can It Possibly Be So Weird? 187
Geneaology of Mixed States 188
Hidden Things 192
The EPR Gauntlet: Entanglement 195
What Bohr Said to EPR 200
A Deeper Theory? 202
John Bell 204
Bell's Theorem Unveiled 205
Bell's Thought Experiment in English (Almost) 209
Nonlocality and Hidden Variables 214
What Kind of World Is This, Anyway? 215

CHAPTER 8: Modern Quantum Physics 219
Marrying Quantum Physics to Special Relativity 221
$E = mc^2$ 221
The Century of the Square Root 223
Paul Dirac 225
Fishing the Dirac Sea 228
The Trouble with the Energy of the Dirac Sea 232
Supersymmetry 235
Holography 237
Feynman's Sum over Paths 238
Condensed Matter Physics 242
The Conduction Band 242
Diodes and Transistors 245
Profitable Applications 246

CHAPTER 9: Gravity and Quantum Theory: Strings 249
General Relativity 251
Black Holes 255
Quantum Gravity? 257

String Theory 258
Superstring Theory 263
Strings Today 265
The Landscape 268

CHAPTER 10: Quantum Physics for Millennium III 271
So Many Worlds . . . So Little Time 275
To Be and Not to Be 277
Quantum Wealth 278
Quantum Cryptography 279
Quantum Computers 281
Future Wonder Computers 283
Finale 286

APPENDIX: SPIN 289
What Is Spin? 289
Exchange Symmetry 293
Bosons 295
Fermions 296

NOTES 299

INDEX 329

INDEX OF FIGURES 337

ACKNOWLEDGMENTS

We thank our editor Linda Greenspan Regan and the production staff at Prometheus Books for their tireless efforts on our behalf to bring this book into production. For useful comments and advice, we thank Ronald Ford and William McDaniel, and for artwork, we thank Ilse Lund and Shea Ferrell.

The authors also underscore the importance of the learning of science by young people. Here, the continuing efforts of Prometheus Books in the publication of science books is gratefully acknowledged, as is the huge contributions of schools throughout the country. We especially acknowledge the Illinois Mathematics and Science Academy, recognized as one of the most successful science high schools in the nation.

Chapter 1

If You're Not Shocked, You Haven't Understood

In the TV series *Star Trek*, and in its subsequent derivatives, the starship *Enterprise* travels throughout intergalactic space. Its five-year mission of exploration is to go where no human being has gone before. Using the imaginative technology of the distant future, the crew of the *Enterprise* travels at warp speeds, many times the speed of light, calls home to Star Fleet Command from a distance of many parsecs, using "subspace communication," and scans approaching vessels and the surfaces of new planets, occasionally defending itself against hostile forces with photon torpedoes. And, perhaps most innovative of all, the starship crew members can "beam" themselves to the surfaces of many new worlds to explore strange landscapes and have face-to-face meetings with the leaders of alien civilizations, which are sometimes more, sometimes less, advanced.

In not one of the many episodes of *Star Trek*, however, or to our knowledge any other science fiction saga, has there ever been as bizarre an exploration of the universe as that which actually took place on planet Earth in the period 1900 to 1930 CE. The distances traveled by the explorers of the early twentieth-century scientific age were similarly vast, but not in the sense of the large scales of billions and billions of light-years of intergalactic space. Rather, it was a voyage into the deep, the unknown, and the unexplored space of the smallest objects that make up the entire universe, down to the scale of billionths and billionths of an inch.

The advancing technology and scientific skills at the turn of the nineteenth to the twentieth century enabled these scientist explorers to, in a sense, visit for the first time the domain of a remarkable and new alien civilization, the world of the atom. What they encountered was incredible, existential, and surreal: it was as if the art, music, and literature of the age—the eyes of Picasso, the ears of Schoenberg, and the pen of Kafka—were in lockstep with the physicists unraveling a weird, bizarre, and unfamiliar new world within the innermost depths of nature. Virtually all of science's sophisticated and well-honed "classical" knowledge of the laws of physics, with its rules acquired and polished over the previous three hundred years, proved to be dead wrong in this strange new world. It was as if Captain Kirk and his *Enterprise* mates had landed on a planet in which the very laws of nature were as different as those encountered by Alice after she fell down the rabbit hole. It was a new kind of "dream logic" reality. Objects placed over here appeared over there, instantaneously. A smooth, hard stone began to blur and diffuse into seeming nothingness as scientists watched. Solid walls could be promenaded straight through, effortlessly. Things jumped wildly about in space and time.

Plenty of "particles" of matter existed in this strange new world, swarming around, to and fro. By carefully observing these particles the scientists learned that they did not simply pass uniformly from starting point A to arrive at a well-defined time at destination point B. Motion was nothing as Galileo or Newton had conceived it three hundred years earlier.[1] Instead, the "fundamental particles" of nature, out of which everything is composed, such as the tiny electron, were seen to explore *all possible paths* in getting from A to B—all at once! Particles were always nowhere and yet everywhere at the same time. They arrived at their destinations with a spooky knowledge of every available path they could have, or might have, taken, with no certainty as to which path they actually did take. The scientists toyed with the particles, blocking off some of the available paths they might have taken from A to B, and they found that their arrival at B could be influenced in this way—merely changing one of the many available paths a particle may have taken, whether it did so or not, could cause it to arrive at B more often—or not at all.

Particles, little pinpoints of matter with no apparent or discernable internal clockwork or organs, leave sharp tracks in detectors and little dots of light on fluorescent screens and cause Geiger counters to go "click . . . click . . . click, click . . . click." Yet, these minute dots of matter now also appear to be waves. They display wavelike, cloudlike, blurry patterns of motion, with crests and troughs like the waves on the surface of a lake or the sea. And things that were thought to be waves, like radio waves and light, were now found to be particles. Waves became particles and particles became waves. Neither, or, yet both, and all at once. It was as if the radical artists, composers, and writers of the age were scripting the laws of nature.

In short, the world dramatically changed before the eyes of the early twentieth-century explorers—eyes that now peered through highly sophisticated instruments. The universe was now seen to work in a way starkly different from what science had taught over the previous three centuries of enlightenment, beginning with the Renaissance. This grand change of our understanding of the physical world marked the arrival of an entirely different way to view nature and was now giving rise to the birth of a whole new and more fundamental science—quantum physics.

Physicists wrestling with the new experimental data and theoretical ideas about the atom strained to use human language and metaphors that had been invented in the traditional world of the old classical era of Galileo and Newton, but they found them hopelessly inadequate to describe their new experiences. The world now seemed to require descriptors such as *fuzzy*, *uncertain*, and *spooky action-at-a-distance*, as if ghosts were running around influencing the outcome of experiments.

There emerged a new concept called "wave particle duality" to reconcile why waves were sometimes particles and why particles were sometimes waves, though scientists were still bewildered. So bizarre are the consequences of quantum physics that, perhaps to preserve their sanity, the quantum physicist pioneers were driven to denial that they were actually describing a vast new reality, preferring to objectively insist that they had "merely" invented a new method for making predictions about the results of possible experiments—and nothing more.

AN ADVANCE PEEK AT QUANTUM PHYSICS

Prior to the quantum era, scientists had been definitive in their statements about cause and effect and precisely how objects move along well-defined paths, as they respond to various forces applied to them. But the classical science that had evolved from the mists of history, to the end of the nineteenth century, always involved descriptions of things that are collections of a huge number of atoms. For instance, some million-trillion atoms are contained in a single grain of sand.

Observers prior to the quantum age were like a distant alien civilization examining large collections of humans from afar, observing them only in crowds of thousands, tens of thousands, or more. They might have observed humans marching in parades, or breaking into applause, or scurrying to work, or about in every which way. Nothing would have prepared such remote alien observers for what they would find upon examining individual human behavior up close. New behaviors would then be encountered as humans displayed signs of humor, love, compassion, and creativity, traits that would be totally unexpected, given the experience of having only observed the behavior of human mobs from afar. The aliens, if they themselves were insects or automatons, may not even have a ready vocabulary to describe what they were now observing in up-close human behavior—indeed, even today the accumulated poetry and literature of the human race, for example, from Aeschylus to Thomas Pynchon, does not span all individual human experience.

Likewise, at the beginning of the twentieth century, the exacting edifice of physics with its precise predictions for the behavior of objects filled with huge orchestras of atoms, all crashed down to the floor. Through newly refined and highly sophisticated experimentation, the properties of individual atoms, and even the smaller particles that are contained within them, now came onto the stage, performing solo or in small ensembles, one's, two's, three's and more. The observed behavior of the individual atom was shocking to the leading scientists, who were awakening from the classical world. These new world explorers, the avant garde "poets, artists, and composers" of the modern physics of the atomic epoch, included such luminary figures as Heinrich Hertz, Ernest Rutherford, J. J. Thompson, Niels Bohr, Marie Curie, Werner Heisenberg, Erwin Schrödinger, Paul Dirac, Louis-Victor de Broglie, Albert

Einstein, Max Born, Max Planck, Wolfgang Pauli, among others. This group was as shocked by what they found inside the atom as the voyagers of the starship *Enterprise* would have been in encountering any alien civilization across the vast reaches of the universe. The new confusion produced by the earliest data gradually gave way to desperate efforts of these scientists to restore order and logic to this new world. Still, by the end of the 1920s, the basic logic of the properties of the atom, which define all of chemistry and everyday matter, was finally constructed. Humans had begun to comprehend this bizarre new quantum world.

However, whereas the *Star Trek* explorers could beam up and ultimately return to less threatening spaces, the physicists of the early 1900s knew that the weird new quantum laws that ruled the atom were primary and fundamental to everything—everywhere in the universe. Since we are all made of atoms, we cannot escape the implications of the reality of the atomic domain. We have seen the alien world, and it is us!

The shocking implications of the new quantum discoveries unnerved many of the scientists who discovered them. Much like political revolutions, the quantum theory mentally consumed many of its early leaders. Their béte noire was not the political machinations and conspiracies of others, but rather deep, unsettling new philosophical problems about reality. When the full force of the conceptual revolution emerged toward the end of the 1920s, many of the originators of the quantum theory, including no less than Albert Einstein, rebuked and turned away from the theory they had a significant hand in creating. Yet, as we plunge into the twenty-first century, we now have a quantum theory that works in every situation we have applied it to, that has delivered to us transistors, lasers, nuclear power, and countless other inventions and insights. There are still strenuous attempts by distinguished physicists to find a kinder, gentler understanding of the quantum theory, less disruptive to the comfort zone of human intuition. But we must come down to dealing with science, not bromides.

The prevailing science, before quantum theory, had successfully accounted for the world of the large: the world of ladders propped safely against walls, the flight of arrows and artillery, the spinning and orbiting of planets and itinerant comets, the world of functioning and useful steam engines, telegraphy, electric motors and generators, and radio broadcasting. In short, almost all the phenomena that scientists could easily observe and measure by the year 1900 were successfully explained by this *classical physics*. The attempt to accommodate the weird behavior of atomic-sized things was enormously difficult and philosophically jarring. The emergent new quantum theory was totally counterintuitive.

Intuition is based on previous human experience, but even in this sense, most of the earlier classical science was itself counterintuitive to the contemporaries of its discoveries. Galileo's insight into the motion of bodies in the absence of friction was extremely counterintuitive in its day (few people had ever experienced or considered a world without friction).[2] But the classical science that emanated from Galileo redefined our own sense of intuition for the three hundred years leading up to 1900, and it seemed impervious to any radical changes. That was until the discoveries of quantum physics brought on an entirely new level of counterintuitive and existential shock.

To understand the atom, to create a synthesis of the apparently self-contradictory phenomena that came out of the laboratories in the period of 1900–1930, meant that attitudes and disciplines had to be radicalized. Equations, which on the large scale made sharp predictions about events, now yielded only possibilities, and to each possibility one could now compute only a "probability"—the probability of the actual physical occurrence of an event. Newton's equations of absolute exactitude and certainty ("classical determinism") were replaced by Schrödinger's new equations and Heisenberg's mathematics of fuzziness, indeterminacy, and probability.

How does this indeterminacy exhibit itself in nature at the atomic level? It does so in many places, but a simple example can be given here. In the lab we learn that if we have a collection of some radioactive atoms, such as uranium, half of the number of atoms will disappear (we say that they "decay into other atomic fragments") within a certain interval of time, called the "half-life." After another period of time equal to the half-life, the remaining atoms are again reduced by half (so, after two half-life intervals we have only one quarter of the original number of radioactive atoms left; after three half-lives we have only one eighth the original number; and so on). We can, in principle, with enough effort and using quantum physics, calculate the half-life of uranium atoms. Many other atomic decay half-lives can similarly be computed for fundamental particles, which keeps atomic, nuclear, and particle physicists gainfully employed. Quantum theory, however, *simply cannot predict when any one uranium atom will disappear.*

This is a jarring result. If uranium atoms were truly governed by Newtonian classical laws of physics, then there would be some internal mechanism at work that, with sufficient detail of study, would allow us to predict exactly when a particular atom would decay. The quantum laws are not just blind to such an internal mechanism, giving us only a fuzzy probabilistic result out of mere ignorance. Rather, quantum theory asserts that probability is all one can ever possibly know about the decay of a particular atom.

Let's consider another example of this quantum aspect of the world: if two precisely identical photons (the particles that make up light) are aimed in exactly the same way at a glass window, one or both photons may penetrate the glass or one or both may reflect back. The new quantum physics *cannot* predict exactly which one of the photons will do which—reflect or penetrate. We cannot, *even in principle*, know the exact future of a particular photon. We can only compute the probabilities of the various possibilities. We may compute, using quantum physics, that "each photon has a 10 percent probability of reflecting off the glass and a 90 percent probability of being transmitted through the glass." But that's all. Quantum physics, despite its apparent vagueness and inexactitude, provides a correct procedure, in fact, the only correct procedure, for understanding how things work. It also provides the only way to understand atomic structure, atomic processes, molecule formation, and the emission of radiation (all light we see comes from atoms). In later years it proved to be equally successful in understanding the nuclei of atoms, how quarks are eternally bound together inside the protons and neutrons of the atomic nucleus, and how the Sun generates its enormous energy output.

How, then, does the classical physics of Galileo and Newton, which dramatically fails to describe the atom, so elegantly predict exactly when solar eclipses will occur, the return of Halley's comet in 2061 (Thursday afternoon), and the exact trajectories of space vehicles? We all depend on the success of Newtonian physics to ensure that airplane wings stay attached and can fly, or that bridges and skyscrapers remain stable in the wind, or that robotic surgical tools are accurate and precise. Why does it all work so well if quantum theory says emphatically that the world really doesn't work this way after all?

It turns out that when huge numbers of atoms congregate together into large objects, as they do in all the above examples of wings and bridges, and even robotic tools, then the spooky, counterintuitive quantum behaviors—loaded as they are with chance and uncertainty—average out to the apparent proper and precise predictability of classical Newtonian physics. The short answer is that it's statistical. It's a bit like the statistically exact statement that the average American household has 2.637, residents, which can be a fairly precise and accurate statement, even though not a single household has 2.637 residents.

In the modern world of the twenty-first century, quantum physics has become the staple of all atomic and subatomic research, as well as much material science research and cosmic research. Many trillions dollars a year are generated in the US economy by exploiting the fruits of quantum theory in electronics and other areas, and many trillions more dollars are generated due to the efficiency of productivity that an understanding of the quantum world has brought forth. A few mavericks, however—physicists who are cheered on by the existentialist philosophers—are still working on the foundational ideas that define quantum theory, trying somehow to make sense of it all, hoping, perhaps, that there is a deeper inner exactitude within quantum theory that has somehow been missed. But they are in the minority.

WHY IS QUANTUM THEORY PSYCHOLOGICALLY DISTURBING?

Albert Einstein famously said: "You believe in a God that plays dice, and I in complete law and order in a world where objectivity exists, and which I, in a wildly speculative way, am trying to capture. . . . Even the great initial success of the quantum theory does not make me believe in a fundamental dice game, although I am well aware that your younger colleagues interpret this as a consequence of [my] senility."[3] And Erwin Schrödinger lamented: "Had I known that my wave equation would be put to such use, I would have burned the papers before publishing. . . . I don't like it, and I'm sorry I ever had anything to do with it."[4] What disturbed these eminent physicists, who turned away from their own beautiful babies? Let's examine the above complaints of Einstein and Schrödinger, often summed up as quantum theory, revealing that "God plays dice with the universe." The breakthrough that led to modern quantum theory came in 1925 when a young German, Werner Heisenberg, on a lonely vacation to Helgoland—a small island in the North Sea where the German scientist sought relief from severe hay fever—had his big idea.[5]

A new hypothesis was garnering strength in the scientific community, that atoms were composed of a dense central nucleus with electrons orbiting around it, like planets orbiting the Sun. Heisenberg pondered the behavior of the electrons in atoms and realized that he did not require any knowledge of the precise orbital paths of the electrons around the nucleus. The electrons seemed to undergo mysterious jumps from one orbit to another that were accompanied by the emission of light of a very precise and definite color (colors are the "frequencies" of the emitted wave of light). Heisenberg could make some mathematical sense of this, but he did not need a mental picture of an atom as a tiny solar system with electrons moving in definite orbits to do so. He finally gave up trying to compute the path of an electron if it was released at point A and detected at point B. And Heisenberg understood that any measurements made on the electron between A and B would necessarily disturb any hypothetical path the electron was on. Heisenberg developed a theory that gave precise results for emitted light from atoms, but that didn't require knowledge of the path taken by the electrons. He saw that, ultimately, only possibilities for events and their probabilities of occurring, with intrinsic uncertainties, exist. This was the emerging new reality of quantum physics.

Heisenberg's revolutionary solution to the results of a set of baffling experiments on atomic physics freed up the thinking of his predecessor, Niels Bohr, the father, grandfather, and midwife of quantum theory. Bohr carried Heisenberg's radical ideas a major step forward, so much so that even Heisenberg was shocked. Ultimately he recovered to join in Bohr's zealotry, while many of his elder and distinguished colleagues did not. What Bohr argued was that if knowledge of the particular path of an electron was not relevant to determine atomic behavior, then the very idea of a particular, well-defined electron "orbit," like that of a planet going around a star, must be a meaningless concept and must therefore be abandoned. *Observation and measurement is the ultimate defining activity; the act of measurement itself forces a system to choose one of its various possibilities*. In other words, reality is not merely disguised by the fuzziness of an uncertain measurement—rather, it is wrong to even think about reality as yielding certainty in the conventional Galilean sense when one arrives at the atomic level of nature.

In quantum physics, there appears to be an eerie connection between the physical state of a system and conscious awareness of it by some observing being. But it's really the act of measurement by any other system that resets, or "collapses," the quantum state into one of its myriad possibilities. We'll see just how spooky this is when we consider electrons, one at a time, passing through one of two holes in a screen, and how the observed pattern of the electrons, detected far from the screen, depends on whether anyone or anything knows which hole the electron went through (or didn't go through), in other words, whether a "measurement" of which hole the electron passed through has been made. If so, we get a certain result. But if not, we get an entirely different result. The electrons seem eerily to take both paths at once if nothing is watching, but a definite path if someone or something is watching! These are not particles and not waves—they are both and neither—they are something new: They are *quantum states*.[6]

Small wonder that so many of the physicists who had participated in the formative phases of atomic science could not accept these strange occurrences. The best face to put on the Heisenberg/Bohr interpretation of quantum reality, sometimes called the "Copenhagen Interpretation," is that when we measure something in the atomic domain, we introduce a major interference into the state itself, in other words, the measuring instruments. In the end, however, quantum physics simply doesn't correspond to our innate sense of reality. We must learn to get used to quantum theory by playing with it and testing it, by doing experiments and setting up theoretical problems that exemplify various situations, getting familiar with it as we go along. In so doing, we develop a new "quantum intuition," as counterintuitive as it all may initially feel.

Another major quantum breakthrough, totally independent of Heisenberg's, took place in 1925 by another theoretical physicist, also on vacation but not quite as lonely. Erwin Schrödinger, the Viennese-born theorist, formed one of the most famous scientific collaborations, with his friend, the physicist Herman Weyl. Weyl was a powerful mathematician who was instrumental in the development of relativity and the relativistic theory of the electron. Weyl helped Schrödinger with his math, and in return, Schrödinger allowed Herman to sleep with his wife, Anny. We don't know how Anny felt about this, but such experiments in marital relationships were not uncommon among intellectuals of late Victorian Viennese society. This arrangement had the collateral benefit of allowing Schrödinger the freedom to indulge in copious extramarital affairs, one of which led (sort of) to one of the great discoveries in quantum theory.[7]

In December 1925, Schrödinger departed for a two-and-a-half week vacation to a villa in the Swiss Alpine town of Arosa. He left Anny at home and was accompanied by an old Viennese girlfriend. He also took with him some scientific papers, by the French physicist Louis de Broglie, and two pearls. Placing a pearl in each ear to screen out any distracting noise, and poring over de Broglie's papers (we don't know what the lady friend was doing), Schrödinger invented "wave mechanics." Wave mechanics was a novel way to understand the nascent quantum theory in terms of simpler mathematics, equations that were already essentially familiar to most of the physicists of the day. This breakthrough significantly promoted the fledgling science of quantum physics to a much broader audience of physicists.[8] Schrödinger's now-famous wave equation, often called the "Schrödinger equation," may have accelerated the progress of quantum physics, but it drove its founder to distraction because of its eventual interpretation. It is astonishing that Schrödinger later regretted publishing this work due to the revolution in thought and philosophy that it inspired.

What Schrödinger did was to describe the electron—in mathematical terms—as a wave. The electron, previously thought to be a hard little ball, indeed also behaves like a wave in certain experiments. Wave phenomena are familiar to physicists. There are countless examples—water, light, sound in air and solids, radio, microwaves, and so on. These were all well understood by physicists in those days. Schrödinger insisted that particles, such as the electron, were actually at the atomic quantum level a new kind of wave—they were "matter waves." As odd as that sounded, his equation was convenient for physicists to use and seemed to come up with all the right answers in a straightforward manner. The wave mechanics of Schrödinger gave a certain comfort to the physics community, the members of which were grappling to understand the burgeoning area of quantum theory and who, perhaps, found Heisenberg's theory too abstract.

The key factor in Schrödinger's equation is the thing that is the solution to the wave equation, which describes the electron wave. It is denoted by the Greek symbol Ψ ("psi"). Ψ is known as "the wave function," and it contains all we know or can know about the electron. When one solves the equation, it gives Ψ, as a function of space and time; in other words, Schrödinger's equation tells us how the wave function varies throughout space and how it changes in time.[9]

Schrödinger's equation could be applied to the hydrogen atom and it completely determined exactly what kind of dance the electrons are doing in the atom: the electron waves, described by Ψ, were indeed ringing in various wavelike patterns, much like the ringing patterns of a bell or other musical instruments. It is like plucking the strings of violins or guitars; the resulting vibrations of the matter waves could be assigned a definite and observable shape and a certain amount of energy. The Schrödinger equation thus gave the correct values of the vibrating energy levels of the electrons in an atom. The energy levels of the hydrogen atom had been previously determined by Bohr in his first guess of a quantum theory (now relegated to the term "old quantum theory"). The atom emits light of definite energy—the "spectral lines" of light—and these are now seen to be associated with the electrons hopping from one vibrational wave-state of motion, say, "Ψ_2", to another vibrational state, say, "Ψ_1".

Such was the newfound power of Schrödinger's equation. One could readily visualize the wave patterns by looking at the mathematical form of Ψ. The wave concept could be readily applied to any system requiring a quantum treatment: systems of many electrons, whole atoms, molecules, crystals and metals with moving electrons, protons and neutrons in the nucleus of the atom, and—today—particles composed of quarks, the basic building blocks of protons and neutrons and all nuclear matter.

In Schrödinger's mind, electrons were exclusively waves, like sound or water waves, as if their particle aspect could be forgotten or was illusory. In Schrödinger's interpretation, Ψ was a new kind of wave of matter, plain and simple. But ultimately Schrödinger's interpretation of his own equation turned out to be wrong. While Ψ had to represent some kind of wave, what exactly was this wave? Electrons still behaved, paradoxically, like pinpoint particles, producing pinpoint dots when impacting a fluorescent screen. How was this behavior to be reconciled with the matter wave Ψ?

The German physicist Max Born (grandfather of singer Olivia Newton-John) soon came up with a better interpretation of Schrödinger's matter wave, and it has become the major tenet of the new physics to this day. Born asserted that the wave associated with the electron was a "probability wave."[10] Born said that it is actually the mathematical square of Ψ, that is, Ψ^2, that represents *the probability of finding the electron at a location x in space at time t.* Wherever, in space or time Ψ^2 is large, there is a large chance of finding the electron. Wherever Ψ^2 is small, there is little chance. Where $\Psi^2 = 0$, there is no chance. Like the Heisenberg breakthrough, this was the ultimate shocking idea, yet in the clearer, easier-to-comprehend, Schrödinger picture of things. Everyone now understood it, finally and once and for all.

Born was clearly saying that we don't know—and can't know—exactly where the electron is. Is it here? Well, there's an 85 percent chance that it is. Or is it there? It might be, with a probability of 15 percent. The probability interpretation clearly defined what you could or couldn't exactly predict about any given experiment in the lab. You can do two apparently identical experiments and get two quite different results. Particles appear to have the luxury of where to be and what to do, with no regard to ironclad rules of cause and effect that one normally had associated with classical science. In the new quantum theory, God does in fact play dice with the universe.

While it rattled Schrödinger that he had been a major player in this unsettling revolution, a further irony is that Born was inspired in his probability interpretation by a speculative paper that appeared in 1911 written by, of all people, Albert Einstein. Schrödinger and Einstein would remain an antiquantum tag team for the rest of their lives. So too, Max Planck: "The probabilistic interpretation put forth by the group from Copenhagen must surely go down as a treasonable behavior towards our beloved physics."[11]

The great theoretical physicist of turn-of-the-century Berlin, Max Planck, was upset with the emerging meaning of quantum theory. This was an extraordinarily ironic development when you consider that Planck was the true grandfather of the new theory and that he had even coined the term *quantum* to describe this new science back in the nineteenth century.

We can appreciate why some would consider as treasonous the endorsement of probability as ruling the universe rather than exact cause and effect. Take an ordinary tennis ball and throw it against a smooth concrete wall so that it bounces back to you. Stand in the same spot, and keep hitting the ball with your racket with the same force and toward the same point on the wall. All other things being equal (wind speed, etc.), as you develop your skill, the ball will always come back to you in exactly the same way, time after time, until your arm gets tired or you wear out the ball (or the wall). Andre Agassi depended on such principles to win at Wimbledon, just as Cal Ripken Jr. made his reputation judging the caroms of baseballs off Louisville Sluggers in Camden Yard. But what if you couldn't depend on the bounce? What if, on one occasion, the ball passed right through the cement wall? And what if it was only a matter of percentages? Fifty-five times the ball bounces back to you; forty-five times it passes through the wall! Some fraction of the time it is returned by the tennis racket, while other times, at random, it passes right through the racket. Of course, this never happens when we are dealing with the macroscopic Newtonian world of tennis balls. But the atomic world is different. An electron impinging upon an electronic wall has a finite probability of passing through the wall (a phenomenon called "tunneling"). So you can imagine the quantum tennis with quantum tunneling would be very challenging and very frustrating.

One can see the probabilistic behavior of photons in ordinary everyday occurrences. Suppose you look at a window display at your favorite Victoria's Secret lingerie store. Superimposed over the shoes of the sexy mannequins you observe a faint image of yourself in the window. What is happening? Light is a stream of particles—photons—that produce a bizarre quantumlike result. Most of the photons—say, coming from a source like the Sun—reflect off your face and pass right through the store window, providing a clear image of you (handsome devil!) to anyone who happens to be on the other side of the window (the window mannequin dresser?). But some small fraction of the photons are reflected back from the glass to provide that dim image of you overlaying the skimpy undergarments in the window display. All photons are identical, so why are some transmitted and others reflected?

In careful experiments it becomes clear that there is no way of predicting which photons will be transmitted and which will be reflected. Only the probability of transmission or reflection can be computed for any given photon. Applying quantum theory to a photon headed for the store window, the Schrödinger equation might tell us that 96 percent of the time the photon will pass through the glass, and 4 percent of the time it will be reflected. But which photon does what? It can't be determined even with the best instruments one could ever imagine building. God rolls dice to find out, or so goes the quantum theory (well, maybe it's a roulette wheel, but it's all about probability, whatever He or She is using).

You can duplicate the store window experience—at much greater expense—by firing electrons at an electric barrier, consisting of a wire screen in a vacuum connected to the negative battery terminal with a potential of, say, ten volts. If the electrons have an energy of only nine volts, they should be repelled, in other words, "reflected." Nine volts of electron energy is not enough to overcome the ten-volt repulsive force of the barrier. But Schrödinger's equation shows that some part of the electron wave penetrates and some is reflected, just like light quanta through the store window. However, we never see a fraction of an electron or a fraction of a photon. These particles do not come apart like a wad of Silly Putty. The particle is always either completely reflected or transmitted. A 20 percent reflectivity means that there is a probability of 20 percent that the entire electron is reflected. Schrödinger's equation gives us the solution, in terms of Ψ^2.

It was just this type of experiment that led the physics world to abandon Schrödinger's interpretation of Silly Putty–like electrons as matter waves and accept the more bizarre idea that Ψ, a mathematical wave function, when squared, describes the probability of finding the electron somewhere. If we fire, say 1,000 electrons at a screen, a Geiger counter might indicate that 568 penetrate the screen and 432 are reflected back. Which ones will do which? We don't know. We can't know. That's the maddening truth about quantum physics. Only the probabilistic odds, Ψ^2, can be calculated.

SPOOKY ACTIONS AT A DISTANCE

Albert Einstein further declared: "I cannot seriously believe [in the quantum theory] because it cannot be reconciled with the idea that physics should represent a reality in time and space, free from spooky actions at a distance."[12]

Einstein thought he had found a fatal flaw in one of the basic principles of quantum physics. Its proponents (especially Bohr) insisted that a particle's various attributes have no real objective reality until they are measured. To Einstein, it was nonsense that objects do not exist until we measure them. To him, particles exist, and have properties like position, velocity, mass, and charge, even when we don't observe them, even when we are ignorant of what values they have. He agreed only with the commonsense idea that measurement of a tiny particle can disturb it and introduce unknown changes. The notion that a quantum state can abruptly change (as in being "reset" as described in chapter 7, note 8) throughout the entire universe, by merely observing it, conjured a notion that signals (information) were somehow being transmitted instantaneously over vast distances, faster than the speed of light, which is impossible. Nothing can exceed the speed of light, according to Einstein's own theory of relativity.

So in 1935, Einstein described a thought experiment that would put an end to the notion that reality is only "enforced" or comes into being by measurement. It was called the EPR Thought Experiment, after Einstein and his two collaborators, Boris Podolsky and Nathan Rosen. The EPR Thought Experiment examined the case of two particles that are produced from the radioactive disintegration of a "parent particle," and whose properties, velocities, spins, electric charges, and so on, are correlated. For example, suppose an electrically neutral radioactive "parent particle" disintegrates somewhere in distant space into two "daughter particles," an electron of negative charge (called Molly) and a positron of positive charge (called June). The two particles stream off in opposite directions with equal and opposite electric charges, but we don't know which one goes which way. June may go to, say, Peoria, and Molly to the distant star Alpha Centauri. But it could be reversed, with Molly heading to Peoria, and June to Alpha Centauri. In classical physics, it is one or the other. But in quantum theory, the actual physical *quantum state* can be described as an indefinite mixture of both possibilities, an "entangled state":

(June→Peoria, Molly→α Centauri) + (June→α Centauri, Molly→Peoria)

This property of adding together, or "superimposing," two (or more) definite possibilities, to create a "mixed" or "entangled" state," is characteristic of the quantum theory that enjoys the privilege of encompassing all possibilities at the same time.[13] Which of the two possibilities corresponds to reality is simply unknown until a definite measurement is made, at which point the quantum state instantaneously changes to reflect the result of that measurement.

So, here's the weird part: When we measure the electric charge arriving in Peoria, we instantaneously learn, without measuring it, the electric charge that arrived at Alpha Centauri, way out there en route to the stars, instantaneously. That is, if we observe June in Peoria, we immediately learn that it is Molly who arrived at Alpha Centauri. The quantum state then immediately changes, or "collapses," *throughout the entire universe*, to become a "pure state":

(June→Peoria, Molly→α Centauri)

It could be the other way around, (Molly→Peoria, June→ α Centauri), as the quantum theory carries both options and merely predicts the probability for either option.

Some might argue that if classical physics were true, we would learn the same thing. But classical physics doesn't require any change in the *state* of nature upon making this observation—we just happen to learn what that classical state of nature actually is. Classical states are never mixed states and they always have a definite reality. In observing a classical state only our own personal ignorance has changed. However, in the quantum theory, once we make the measurement, the actual physical state—or wave function—of June and Molly is suddenly changed, instantaneously and everywhere throughout the whole universe, and is now a new quantum state.

To Einstein, this required a seeming instantaneous propagation of information throughout the universe, at least from Peoria to Alpha Centauri, in violation of the ultimate speed limit in nature, the speed of light. Einstein must have exclaimed to Bohr: "Gotcha!"

This, indeed, confronted Bohr with what seemed to be a devastating refutation to his interpretation of the quantum state. Since the location of June can be deduced without measuring the electron Molly, whose properties are correlated by the initial quantum state of the radioactive parent particle, the properties of the particle arriving at Alpha Centauri must seemingly have an objective reality. However, Bohr insisted that a definite property does not exist until a measurement brings it into being. Einstein's conclusion was that quantum theory, in determining the properties way over there by measurements here, implied some "spooky actions at a distance" and that Bohr's interpretation implied signals traveling faster than the speed of light, ergo the quantum theory is incomplete or flawed. This is the kind of problem that caused such physicists as Planck, de Broglie, Schrödinger, and Einstein to reject the form that quantum theory had taken.

Did the EPR Thought Experiment put a silver stake through the heart of the quantum theory? Obviously not! It's still around, alive and well and arguably the most successful theory in the history of science. How did the quantum champions counter Einstein's powerful argument? Essentially, the skinny on this is that "yes, the state indeed resets," or "collapses" instantaneously throughout the entire universe, to one of the two possibilities, but try as you might, there is no experiment that can ever reveal any consequence of a spooky action at a distance. No message can be transmitted to Alpha Centauri faster than the speed of light. The observer there doesn't know that Molly is arriving until he observes who arrived. Likewise, the observer is unaware of the collapse of the entangled quantum state into one of its definite possibilities until he does the measurement. Ergo, the EPR Thought Experiment does not violate the "causality of nature," by which signals travel at speeds less than or equal to the speed of light. Reality, so said Bohr, is still conditioned by the act of measurement. So added Bohr: "Anyone who is not shocked by the quantum theory must not have understood."[14]

A saving grace was that the EPR headache seemed confined to the remote and obscure domain of the atom, in which Newtonian laws had been revoked. But not for long. After all, we are all made of atoms.

SCHRÖDINGER'S TABBY

We can't leave the domain of quantum philosophical hand-wringing without a brief look at the now-famous Paradox of Schrödinger's Cat, which links the squishy quantum microworld and its statistical probabilities to the Newtonian macroworld with its exact statements. Like Einstein, Podolsky, and Rosen, Schrödinger took issue with a world that had no objective reality until measured, one that was just a roiling mass of possibilities, until observed. Schrödinger's paradox was intended as a derisive comment on that worldview, but it's turned out to be one that tantalizes scientists to this day. He figured out a way, by using a *thought experiment*, to make the quantum effects of the atom dramatically apparent in the ordinary macroscopic world. To help him make his case he again enlisted the phenomenon of radioactivity, in which particles decay at a predictable rate, though one cannot predict when any given particle will decay. That is, one can predict what percentage of atoms will decay in, say, an hour's time, but one cannot predict which of the individual atoms will decay.

So here's Schrödinger's recipe: Put a cat in a box, with a flask of lethal gas. In a Geiger tube, place a small quantity of radioactive material, such a minute amount that in the course of an hour we have only a 50 percent chance of detecting a single atomic disintegration. Arrange a "Rube Goldberg" device where the decaying atom will set off the Geiger counter, which will activate a relay that will trigger a hammer, that shatters the flask of gas, that will kill the cat. (Oh, those turn-of-the-century Viennese intellectuals . . .)

So, after one hour, is the cat dead or alive? If one uses the quantum wave function to describe the entire system, the living and dead cat would be a mixed state, "smeared out (pardon the expression) in equal parts." Ψ, the wave function, would tell us that the situation is a mixture of "cat alive" and "cat dead,"[15] that is, there would be a mixed quantum state of the form $\Psi_{\text{Cat Alive}} + \Psi_{\text{Cat Dead}}$. So, even at the macroscopic level we could only determine the probability of finding the cat alive $(\Psi_{\text{Cat Alive}})^2$ or of finding the cat dead $(\Psi_{\text{Cat Dead}})^2$.

But here's the question: is the reset of the quantum state to "cat alive" or "cat dead" is determined the moment *who* (or *what*) looks in the box? Isn't the cat already in there, nervously looking at the Geiger counter and making the measurement himself? Or the "who crisis" can be extended:

the radioactive decay can be monitored by a computer and the state of the cat at any instant can be printed on a slip of paper in the box. Is the cat certainly either dead or alive when the computer first detects it? At the time the complete printed message says so? When I look at the printout? Or when you look at the printout? Or does it happen when the atomic transition produces a cascade of electrons moving in the Geiger tube sensor that yields the Geiger counter "tick," where we transit from the subatomic back to the macroscopic world? Schrödinger's cat-in-the-box paradox, like the EPR experiment, appears to be a strong argument challenging the basic principles of the new quantum theory. By our intuition we obviously can't have a "mixed" cat, half dead, half alive—or can we?

As we shall later see, experiments show that Schrödinger's macroscopic cat, as a metaphor for a large macroscopic system, may in fact exist in a mixed state; in other words, quantum theory can lead to mixed states on the macroscopic level, so quantum physics triumphs yet again.

Quantum effects can indeed range from the level of the tiny atom to the grand scale of macroscopic systems. Such is the quantum phenomenon of "superconductivity" where, at extremely low temperatures, certain materials become absolutely perfect conductors of electricity. Currents will flow in circuits forever without batteries, and magnets can levitate over rings of superconducting current forever. Such is also the phenomenon of "super-fluids," where a stream of liquid helium can move up and down the walls of a vessel, or pump continuously from a pool through a fountain and back again, continuously, forever, without consuming energy. Such is the phenomenon by which all elementary particles acquire mass—the mysterious "Higgs mechanism." There is no escaping quantum mechanics. We are ultimately all cats in the same box.

It went many years
But at last came a knock
And I thought of the door
With no lock to lock.

I blew out the light,
I tip-toed the floor,
And raised both hands
In prayer to the door.

But the knock came again,
My window was wide;
I climbed on the sill
And descended outside.

Back over the sill
I bade a "Come in"
To whoever the knock
At the door may have been

So at a knock
I emptied my cage
To hide in the world
And alter with age.

Robert Frost, "The Lockless Door"[16]

NO MATH, BUT PERHAPS A FEW NUMBERS

Our purpose here is to bring you some sense of the laws of physics that have been developed to comprehend the spooky microworld of atoms and molecules. We ask of the reader only two small qualifications: a curiosity about the world and a complete mastery of partial differential equations. No, wait! We're kidding. We have been teaching freshman liberal arts students for many years and understand thoroughly the layperson's fear and loathing of math. So, no math—not much, at least—maybe just a tad bit now and then.

What scientists say about the world should be a part of everyone's education. And quantum theory, in particular, is the most seminal change in viewpoint since the early Greeks gave up mythology to initiate the search for a rational understanding of the universe. It vastly expanded the domain of human understanding. Modern scientists, expanding our intellectual horizon, came at a price—the price of accepting quantum theory and a lot of counterintuitive spookiness. Remember, this is largely due to the failure of our old Newtonian language to describe the new world of the atom. But we scientists will do the best we can.

Because quantum theory takes us into the realm of the very, very small, let's use the "powers of ten" scale to simplify the task ahead. Please don't be intimidated by the scientific notation (10^4, etc.), that we will use from time to time. This is simply a method of writing very large or small numbers as powers of ten. Think, for example, of 10^4 as simply a 1 followed by *four* zeroes (which is 10 raised to the fourth power). So 10^4 is just 10,000. Conversely, 10^{-4} is a 1 that is four places to the right of the decimal point, or 0.0001 (or 1 divided by 10,000 which is one ten-thousandth).

In this simple language, some of the scales in nature of length or distance can be expressed as powers of 10 in descending order as:

- *One meter* (about 3 feet) is a typical human dimension: the height of a child, the length of an arm, the length of a marching step.
- *One centimeter*, or 10^{-2} meters (say it "ten-to-the-minus-two meters") is about the size of a thumbnail, a honeybee, or a cashew nut.
- *Ten to the minus four* (10^{-4}) *meters* gets us to the thickness of a pin or an ant's leg. This is still in the domain of classical Newtonian physics.
- *Another plunge to* 10^{-6} *(a millionth) meters* brings us to large molecules found in living cells, such as DNA. Here we begin to see the onset of quantum behavior. This is also near the "wavelength" of visible light.
- *An atom of gold* is 10^{-9} (*a billionth*) *mete*r in diameter. The smallest atom, hydrogen, is 10^{-10} meters in diameter.
- *The nucleus of an atom* is 10^{-15} meters; the diameter of a proton or a neutron is 10^{-16} meters, below which we find the quarks inside the proton. 10^{-19} *meters* is the smallest distance scale the most powerful particle accelerator (the world's most powerful microscope), the Large Hadron Collider (LHC) in Geneva, Switzerland, can directly observe.
- 10^{-35} *meters is the smallest distance scale we believe exists*, at which point quantum effects cause distance itself to lose meaning.

We know from experiment that quantum theory is valid and essential in order to extend our knowledge from atoms, 10^{-9} meters, to atomic nuclei at 10^{-15} meters. In words, the size of an atomic nucleus is a thousandth of a trillionth of a meter. In scientists' most recent explorations of the small, we have probed, using the Fermilab Tevatron, to 10^{-18} meters with no hint that the quantum theory doesn't work. Soon, we scientists will descend another factor of ten to smaller distances, as the CERN Large Hadron Collider (LHC) begins to operate. This new territory of the very small isn't just an adjacent neighbor to our everyday world, as was the case for the Europeans with the discovery of America. Rather, it *is* our world, since the universe is made of the all the denizens of the subnuclear world. Its properties, its heritage, and its future are determined thereby.

WHY DO WE CARE ABOUT A "THEORY"?

Some may ask, why should we care about quantum theory if, in fact, it is only a theory? There are theories and then there are theories. We scientists are at fault for using the word *theory* in many different senses. *Theory* really isn't a scientifically well-defined word at all.

Let's take a silly example. People living near the Atlantic Ocean may notice that the sun first appears over the ocean every morning about 5 a.m. and it sets in the opposite direction every evening at about 7 p.m. To explain this, one venerable professor offers up the theory that there are an infinite number of suns lined up just beyond the horizon, spaced twenty-four hours apart. These suns keep popping up on one side of the planet and drop out of sight at the opposite side. A more economical theory is that there is only one sun and it rotates around a spherical Earth, taking twenty-four hours to do so. A third theory, more bizarre and counterintuitive, is that there is a stationary sun, and Earth spins around an axis in twenty-four hours. So we have three competing theories. Here the word *theory* implies a hypothesis or hypothetical idea designed to understand data in some organized, rational way.

The first theory is rather quickly discarded for a variety of reasons. Perhaps the pattern of sunspots is the same every day, or maybe it's just a dumb theory. The second theory is harder to dispose of, but observations of other planets show that they spin about axes, so why not Earth? Eventually, detailed measurements of things near the surface of Earth actually detect our planet's axial rotation, so only one theory survives: the axial rotation theory, or "AR" for short.

Here's the problem: through all this debate we never drop the word *theory* and replace it by the word *fact*. Hundreds of years later, we still refer to the "AR theory," even though it is now as much a fact as anything else we know. What we are saying is that the surviving theory is the one that is best in accord with measurement and observation—the more varied and extreme the circumstances to which the tests are applied, the better. Eventually, it reigns supreme—until a better explanation is proposed. However, we still use the same word, *theory*. Perhaps this can be blamed on the experience that even tried-and-true theories that become facts in a certain domain of application, may eventually require modification as they are extended to larger domains.

So today we have the *theory* of relativity, the quantum *theory*, electromagnetic *theory*, the Darwinian *theory* of evolution, and so on, all of which have now graduated to a higher level of scientific acceptability. These theories all provide valid explanations of their phenomena and are all *factual* in their domains of applicability. We also have new proposed theories—for example, the *theory* of superstrings—that are excellent tentative hypotheses, but that may or may not eventually be established or discarded. And we have old theories, such as "*phlogiston*" (a hypothetical fluid that filled and defined all living beings) and "*caloric*" (hypothetical fluid that was heat), that we have completely discarded. For now, though, quantum theory is the most successful theory in all of science. The fact is: the quantum theory is a fact.

INTUITION? FIRE UP YOUR COUNTERINTUITION

In approaching the new domain of the atom, all our intuition may be suspect. Our prior information may not be relevant. Our normal lives expose us to a very limited range of experiences. We have no experience of traveling at velocities millions of times faster than a speeding bullet. We have not been subject to the searing heat, billions of times hotter than the center of the sun. Neither have we danced with individual molecules, atoms, or nuclei. Whereas our direct experience of nature is limited, science has enabled us to become aware of the vastness and diversity of the world outside of us. A colleague's metaphor has us like an embryonic chick, consuming the stored food inside the egg until it is all gone and the world seems to be at an end. But then the shell is cracked open and the chick emerges into a new and vastly greater (and more interesting) world.

Among the many intuitive ideas that most adults have is that the objects around us—chairs, lamps, cats,—have an existence and a full set of properties whether we are there to observe them or not. Another belief that emerges from our schooling is that if we prepare an experiment on successive days, for example, racing two toy cars down two identical ramps, and conduct them in exactly the same way, we will get the same result. Don't you find it intuitively obvious that if a baseball flies from batter to outfielder, that at each point along its trajectory it has a definite position and a definite speed? A series of snapshots of the baseball (that's what a video is) would serve to locate the ball at any instant. Putting all the snapshots together should define a smooth trajectory.

These intuitions continue to serve us well in the macroscopic world of chairs and baseballs. But, as we have seen and will continue to see, funny things happen inside atoms. Be ready to check some of your most cherished intuitions at the door. The history of science is a history of revolutions that enfold preexisting knowledge. For example, the Newtonian revolution enfolded—rather than discarded—the earlier work and concepts of Galileo, Kepler, and Copernicus, as well as electromagnetic theory, as ultimately summarized by James Clerk Maxwell.[17] Maxwell in the nineteenth century extended and, in part, included Newton's mechanics. Einstein's relativity enfolded Newton's theory and enlarged the domain to include high velocities and a deeper view of space, time, and gravity. Newton's equations remained valid in the domain of small velocities. Quantum theory enfolded the Newton/Maxwell theories so that we could understand the atomic domain. In each case, the newer theory had to be understood in the language of the older theory, at least in the beginning. But when discussing the quantum theory, the language of "classical" theory—our human language—fails.

The problem that Einstein and his fellow dissenters had, and the problem we have today, is the difficulty of comprehending the new physics of the atom in the language and philosophy of the old physics of macroscopic objects. What we must learn to do is understand how the old world of Newton and Maxwell comes about as a consequence of the new stuff, in the language of the quantum theory. If we were atomic-sized scientists, we would have grown up with quantum phenomena. Then some quark-sized alien buddy of ours might say, "What kind of world would we have if we assembled 10^{23} atoms into something we could call a "baseball"?

It may be that probability, uncertainty, objective reality, spookiness, and so on are all concepts that defy our language. This remained a problem even at the end of the twentieth century. Richard Feynman, it is said, refused a TV interviewer's polite request to explain, for the viewing audience, the force between two magnets. "I can't," said the great theoretical physicist. Later he explained why. The TV guy (and most people) understand force as the push of one's hand against the table. This is his world and his language. But the hand on the table involves electricity, involves quantum theory, involves the properties of materials. It's complex. The TV interviewer expected Feynman to explain the pristine, pure magnetic force in terms of forces that would be "familiar" to the denizens of the "old world."

As we shall see, understanding quantum physics means entering a whole new world. It is surely the greatest discovery that scientific exploration has made in the twentieth century and will be essential throughout the twenty-first. It is much too important to leave it only to the pleasure and the profit of professionals.

As we begin life in the second decade of the twenty-first century, there are still strenuous attempts by distinguished physicists to find a more philosophically pleasing, "kinder, gentler" version of quantum theory that is less shocking to human intuitions. But their effort seems to lead nowhere. Other physicists simply master the rules of quantum physics as they are and make strident steps forward, adapting the rules to new symmetry principles, to conjectured strings and membranes instead of pointlike particles, and construct powerful visions of distance scales trillions of times smaller than what we can examine today with our microscopes. This latter approach seems the most productive, giving us a strong hint of a unification of all known forces and the very structure of space and time.

Our intent in this book is to convey both the unsettling creepiness of the quantum theory and its profound consequences for our understanding of nature. We think that much of the eeriness is due to human conditioning. Nature has its language and we must learn it. We should learn to read Camus in French and not force it into American slang. If French gives us difficulties in interpretation, then we should take some long vacations in Provence and breathe in the French air, rather than sit in suburbia and force the world into our own vernacular. In the following pages, we hope to transport you to a world within and beyond our world with the added benefit of acquiring a new language by which to comprehend this brave new world.

Chapter 2

BEFORE THE QUANTUM

When Galileo ascended the Leaning Tower of Pisa to drop two objects of different weights (but shaped identically, so as to equalize air resistance), he was doing more than conducting a scientific experiment. He was creating great street theater, a chance to publicly thumb his nose at the Aristotelian faculty of the University of Pisa. It, perhaps, was also a publicity stunt to stimulate funding (Galileo was forced, at one point in his career, to cast horoscopes for the Medici's to help support himself). More significantly, Galileo was demonstrating the importance of replacing intuition with empirical evidence, and dogma with fact.

As we delve further into quantum theory, your innate intuition about reality and what makes sense about the "physical world" will be challenged mightily. But you may also be no less shocked than were the ordinary people of Pisa who watched and heard those two objects land at the foot of the tower with simultaneous thuds. How could a heavy weight *not* fall to the ground faster than a light one? (How could *Aristotle* be wrong?) Intuition was taught. The ancient Greeks never did the experiment to see which one would hit the ground first. Intuition may not be so innate after all but rather learned by observation.

At the time of Galileo, Europeans had been told for two thousand years (incorrectly so) that heavier objects fall to the ground faster than lighter ones. They were also told that moving objects must naturally eventually come to a halt and that Earth is at the center of the universe with: "all things in their order, Moon, Sun, the planets, circling about, Heaven above and Hell below." Galileo's radical ideas were based on observation and consequential reason—two objects, no matter what their weight (but negating the effects of air resistance), dropped simultaneously, will reach the ground at the same instant. This is a result that can be proved by actually doing the experiment. Moreover, things also move continuously forever in a straight line unless a force acts to change that state of motion, again testable with objects moving on smooth frictionless surfaces. The sun forms the center of a solar system about which the planets (of which Earth is one) move in elliptical orbits, while the moon orbits Earth, thus resolving nagging observational difficulties with the old system of an Earth-centered universe. Galileo's ideas were as "counterintuitive" in 1600 CE as quantum theory was in 1930 CE.[1]

Before one can comprehend the vertiginous universe of quantum physics, one must understand some of the science that preceded it. That science is called *classical physics*, the culmination of hundreds of years of work begun before the time of Galileo and subsequently refined by Isaac Newton, Michael Faraday, James Clerk Maxwell, Heinrich Hertz, and many others.[2] Classical physics posited a kind of clockwork universe: orderly, causal, exact, and predictive, and it reigned supreme until the beginning of the twentieth century.

COMPLICATING FACTORS

To get a sense of a counterintuitive idea, consider Earth, which seems so solid, so eternal, and so stationary. We can comfortably balance our breakfast tray on Earth without spilling a drop of coffee, yet Earth is spinning on its axis. The objects on its surface are not sitting at rest at all but are revolving around with the spinning Earth which acts like a giant merry-go-round. At a surface velocity of up to 1,000 miles per hour near the equator it is as fast as a jet aircraft. Moreover, Earth is streaking through space in its orbit around the sun at a dizzying 100,000 miles per hour. And the whole solar system is hurtling about the galaxy at still higher speeds. Yet, though the sun comes up in the east and sets in the west, we don't feel a thing, and we hardly notice any motion at all. How can this be? It is impossible to write a letter on horseback and even difficult in a car speeding down the interstate at 70 miles per hour, yet we've all seen pictures of astronauts threading a needle and performing other delicate chores in a space capsule orbiting the earth at 18,000 miles per hour. Those floating astronauts, apart from the blue planet turning below, don't appear to be moving at all.

The pattern of the sun
Can fit but him alone
For sheen must have a Disk
To be a sun—

Emily Dickinson, "The Pattern of the Sun"[3]

Our everyday intuition does not immediately inform us that if our surroundings share our same motion, and if the motion is uniform and unaccelerated, we will have no sensation of motion whatsoever. The ancient Greeks believed that there was an absolute state of rest attached to the surface of Earth. Galileo challenged and replaced this time-honored "Aristotelian" intuition with a new and scientifically improved one. Sitting still, we learn, is no different than a uniform, approximately constant, state of motion. The astronauts are sitting still from their point of view, but they are hurtling through space at 18,000 miles per hour from our point of view.

To Galileo's discerning eye it became totally evident that a lighter mass and a heavier mass would fall at the same rate and arrive at the ground simultaneously. To most people, this was not obvious at all, because experience seemed to indicate just the opposite. But Galileo did the experiments that indicated this was so, and he reasoned what it meant. Only the resistance to motion by the surrounding air caused this effect to go unnoticed. In his mind, Galileo saw the surrounding air as a complicating factor obscuring a profound underlying simplicity to nature. He realized that, in the absence of air, all bodies will fall to the ground at the same rate, even feathers and gigantic rocks.

It indeed turns out that the pull of gravity, or the *force* of gravity, does depend on how massive the pulled object is. *Mass* is a measure of the quantity of matter in the object.

The *weight* is just the force of gravity on the massive object (remember what Mr. Jones, your science teacher, always said: "The mass of an object will be the same on the moon, but the weight will be less"—and Mr. Jones, like the rest of us, learned this fact as a result of people like Galileo). The more massive the object, the stronger the force of gravity; double the mass, and you double the force. But it is also true that the more massive the object, the greater is its resistance to changing a state of motion. These two countervailing tendencies exactly cancel each other out so that all objects fall to the ground at the same rate—if we can remove the effect of air resistance. Air resistance is a complicating factor.

It seemed obvious to the ancient Greek philosophers that the most natural state of an object is to be sitting still, at rest. If we kick a soccer ball, it rolls eventually to rest. Your car will keep going only until it runs out of gas, then it slows down and comes to rest. Slide a hockey puck on a lecture table and it goes a few feet and stops. All this is perfectly obvious, all perfectly Aristotelian (we all have an inner Aristotle).

But Galileo developed a deeper intuition: he recognized that if the hockey puck is waxed and polished, it goes farther along the surface of a likewise waxed and polished table; and if the table it slides on is replaced by a frozen lake of ice, it goes a very long distance. Remove all friction and any other complicating factors, and the hockey puck will slide on in a straight line at a constant velocity forever. "Ecco!" reasoned Galileo—what causes the loss of motion is friction between puck and table (or car and road), and that's a complicating factor.

In a typical university lecture room you might find a long steel track punctured with thousands of tiny holes through which air blows. This causes a metal rider (a surrogate hockey puck) to float on air as it moves along the track. The track is terminated on both ends with elastic bumpers. It takes only a slight push to send the rider gliding where it bounces off a bumper, and returns, then bounces again and again, back and forth across the thirty-foot track for the entire lecture hour. Why does it move for so long on its own? It's very entertaining because it is so counterintuitive, but here we witness the underlying primal world, unfettered by the complicating factor of friction. From his more technologically primitive, though equally illuminating, experiments, Galileo discovered and formulated a new law of nature that said: "An isolated body in motion will maintain its motion forever." By "isolated," he meant no forces of friction or anything else. Only forces can change a state of uniform motion.

Counterintuitive? You bet! It is extremely difficult to imagine a truly isolated object, since we never encounter such a beast in the living room, the ballpark, or anywhere else on Earth, for that matter. We can approach this idealized state only in a carefully designed experiment. But after many demonstrations like that of the rider on the air track, this law eventually does become part of the intuition of the average college freshman who studies physics.

The scientific method encompasses careful observation of the world. A vital key to the resounding success of the scientific method over the past four hundred years is that it enables us to abstract, to create a purified mental mini-world, free of real-world complications, and to seek out the basic laws of nature therein. Afterward, we can go back and attack the more complex real world, quantifying such complicating factors as friction or air resistance.

Let's consider another important example: The real solar system is vastly complicated—with a massive star, the sun, at its center and nine smaller planets of varying masses (eight, if you don't count Pluto), each with an assortment of moons, each object pulling on all the other objects and propelling a complex ballet of motion. To simplify things, Isaac Newton posed the simple, idealized question: Consider a solar system in which there is only one planet and one sun. How would they move?

This is called a "reductionist method." You take a complex system (e.g., nine planets and one sun) and you consider a smaller subset of the problem (one planet and one sun). Now, perhaps, the problem becomes solvable (in fact, it does). Then you can find the features that are preserved in the more complicated problem you initially posed (e.g., each one of the nine planets moves mostly like a single planet orbiting the sun, with small corrections from the forces between the planets).

The reductionist method isn't always applicable and it won't always work. This is why tornadoes and the turbulent flow of liquids through pipes are still incompletely understood phenomena today, let alone the complex phenomena of large molecules and living organisms, the most complex of physical systems. The reductionist method works best if the abstract simple system imagined by the physicist is not too different from the real, messy one in which we live. In the case of the solar system, the massive sun dominates all the other planetary forces acting on planet Earth, so we get a rather good answer by ignoring the influence of Mars, Venus, Jupiter, and so on. We get a reasonable description of Earth's orbit by considering just Earth and the sun as a simpler system. Once we gain confidence in our method, we can go back and work harder and include the next most important complicating factors.

THE PARABOLA AND THE PENDULUM

> There was a discordant hum of human voices! There was a loud blast as of many trumpets! There was a harsh grating as of a thousand thunders!
>
> The fiery walls rushed back! An outstretched arm caught my own as I fell fainting into the abyss. It was that of General Lasalle. The French army had entered Toledo. The Inquisition was in the hands of its enemies.
>
> Edgar Allan Poe,
> "The Pit and the Pendulum"[4]

Classical, or prequantum, physics rests on two pillars. The first is seventeenth-century Galilean/Newtonian mechanics. The second consists of the laws of electricity, magnetism, and optics developed in the nineteenth century by a series of physicists whose names bear a curious resemblance to various electrical units: Coulomb, Oerstad, Ohm, Ampere, Faraday, and Maxwell. Let's first consider the universe of the master physicist Isaac Newton, a successor to our hero Galileo.

Objects fall down, and the rate of increase in their speed as they fall has a precise value (this is called *acceleration*). A projectile, a ball batted into the air, or a ball fired from a cannon each soars away from its launching point in a graceful arc of sublime mathematical elegance called a *parabola*. A pendulum, a mass attached to a long string suspended from a great height, as in an old grandfather clock or an old tire tied to a tree branch by a rope, swings to and fro with a precision by which you can set your watch. The sun and moon pull on Earth's oceans to create the tides. All these phenomena are understood and accounted for by Newton's laws of motion.

Newton made two great discoveries in a burst of creativity that has seldom been matched in human history. Both were expressed in a mathematical language called "the Calculus," much of which he had to invent in order to compare his predictions with nature. The first of his discoveries—often presented as Newton's three laws—was a method of calculating the motion of objects once you know the forces acting on them. Newton might have said: "Give me the forces and a big enough computer and I will give you the future." But he didn't, to our knowledge.

The forces that act on an object can be delivered to it by just about anything: ropes, rods, human muscle, wind or water pressure, magnets, and so on. One special force of nature—gravity—became the focus of Newton's second great discovery. In an absurdly simple-looking equation he made the sweeping generalization that all objects attract one another with a force. This force between two objects decreases as the distance between them increases. Double the separation, for example, and the force of gravity decreases by one-fourth. Triple the separation and the force drops to one-ninth. This is the famous "inverse square law." It means that we can go far enough away from an object so that its influence can become as small as we want it to be. The force of gravity we feel from Alpha Centauri, one of the stars nearest our sun—a mere four light-years distant (that is, it takes light four years to get here from Alpha Centauri)—is 1/10,000,000,000,000, or 10^{-13} times our weight on Earth (that fraction of the US annual GDP is about one dollar). Or we can get so close to a dense massive object, like the surface of a neutron star, that we would be squished by gravity into an atomic nucleus. Newton's laws spell out how gravity acts on falling apples, projectiles, pendulums, and other objects near the surface of this planet, where most of us make our living. Gravity also acts across vast stretches of space, across the ninety-three million miles between Earth and the sun, for example.

But do Newton's laws really still work when we leave our native planet? The theory must give results that agree with our measurements (with allowances for experimental error). And—guess what?—the results show that, in general, Newton's laws work well for the entire solar system. In fact, to a very good approximation for each planet we need only consider the simpler case of the single planet orbiting the sun, and Newton's laws then predict perfect elliptical orbits for each planet. But, when we look in greater detail, we find that there are these small discrepancies in the orbital motion of Mars. The orbit of Mars is not a perfect ellipse, as the "two-body" reductionist approximation predicts.

When we analyze, say, the sun-Mars system in isolation, we have left out the relatively tiny gravitational effects of Earth, Venus, Jupiter, and so on, each of which pulls on Mars, too. Mars gets a lot of kicks from Jupiter as they pass each other in their orbits, and, over long time scales, these effects can really add up. Mars might, like a reality TV show participant, even get kicked out of the solar system by Jupiter in a few billion years. So the problem gets more complex as we look at the planetary motion over long periods of time. But with modern computers we can deal with these small (or not-so-small) perturbations, including the tiny effects of Einstein's general theory of relativity (the modern form of gravity theory). After we include all the effects, we find that the theory and observational measurements agree better than ever. But will Newton's laws still operate in the vast, trillion-mile distances between stars? Although the strength diminishes with distance, modern astronomical measurements indicate that the force of gravity reaches out across the cosmos, as far as we can tell, forever.

Now take a moment to contemplate the variety of activities that are beholden to Newton's laws of motion and gravity. Apples fall pretty much straight down, actually falling toward the center of Earth. Artillery shells make their deadly parabolic arcs. The moon hovers a mere 250,000 miles away, tugging at our oceans and our romantic nature. Planets sweep around the sun in elliptical orbits that are nearly circular. Comets zoom in toward the sun and curve around in highly elongated elliptical orbits that may take tens or even hundreds of years before returning. From the smallest to the largest, all the ingredients of the universe move in precisely predictably ways—according to Sir Isaac Newton.

How can one or two simple mathematical equations encompass so many outcomes?

THE CANNON AND THE COSMOS

Newton himself chewed over the problem of the range of scales over which his law of gravity applies. To address it, he devised a hypothetical cannon, perched on the edge of a cliff. In this problem he wanted to calculate the different trajectories of a cannon ball, depending on how much gunpowder was used to shoot it. If we recreated this experiment, we might start with a small bag of cheap, old mildewed gunpowder. The gunpowder would fizzle, barely managing to push the ball out of the muzzle. The cannon ball would roll out of the barrel and fall down the cliff. It would fall almost vertically to the ground, just like an apple falling from a tree—both governed by the force of gravity and the laws of motion.

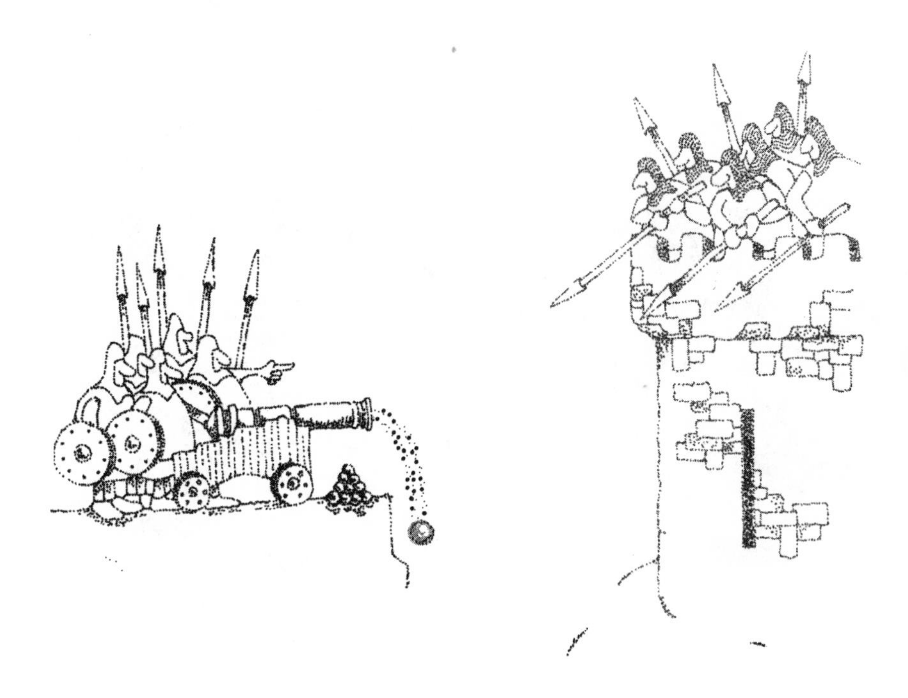

FIGURE 1: The general orders that the cannon be fired with one sack of powder. The powder became damp and mildewed during the long and treacherous march from the East. "Pop"—the cannon fizzles, and the ball is ejected, falling nearly straight to the ground and accelerating downward at $g = 32$ feet/second ^{2}, just like Newton's apple. (Illustration by Ilse Lund.)

So next, we might try a standard bag of fresh government-issue powder. Bang! This time the ball would sail out of the barrel and create a graceful arc, striking the ground a hundred yards from the base of the cliff. It would not be quite far enough for the generals, though. Therefore, we'd better try three bags of powder, and while we were at it, we would elevate the barrel slightly. Pow! The ball would now zoom out of the barrel and make a nice, high parabola, striking the ground five miles away.

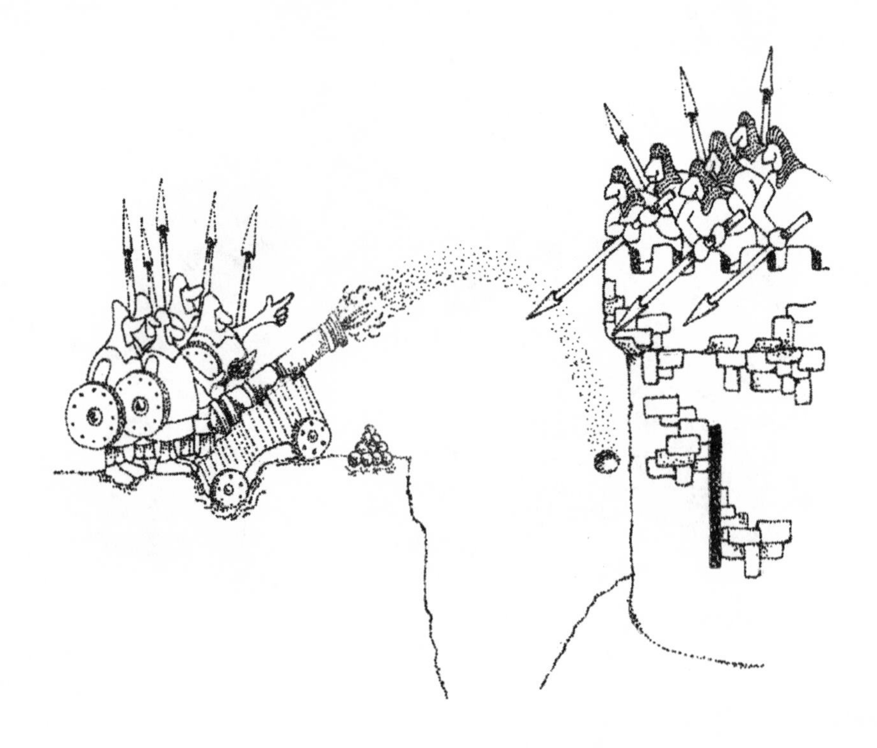

FIGURE 2: The general orders three sacks of the powder. "Boom" and the ball flies toward the castle in a parabolic arc, accelerating to the ground at $g = 32$ feet/second^2 and falling short of the castle walls. (Illustration by Ilse Lund.)

But the generals would still want more bang for their buck, so we would load a special souped-up high-power dynamite—ten bags' worth. This time—KABOOM! The thunderous, flaming explosion would be felt by the generals several miles away in their observation post. Expectantly they would scan the target area but would find nothing. Had the cannon ball disappeared, disintegrated by the very explosion that launched it? They had better phone the boys in the lab. "Ten bags!?" they'd say in utter astonishment, "You idiots put the ball in orbit!" Sure enough, ninety minutes later, the cannon ball would zoom over their heads, having circumnavigated Earth like a new *Sputnik*.

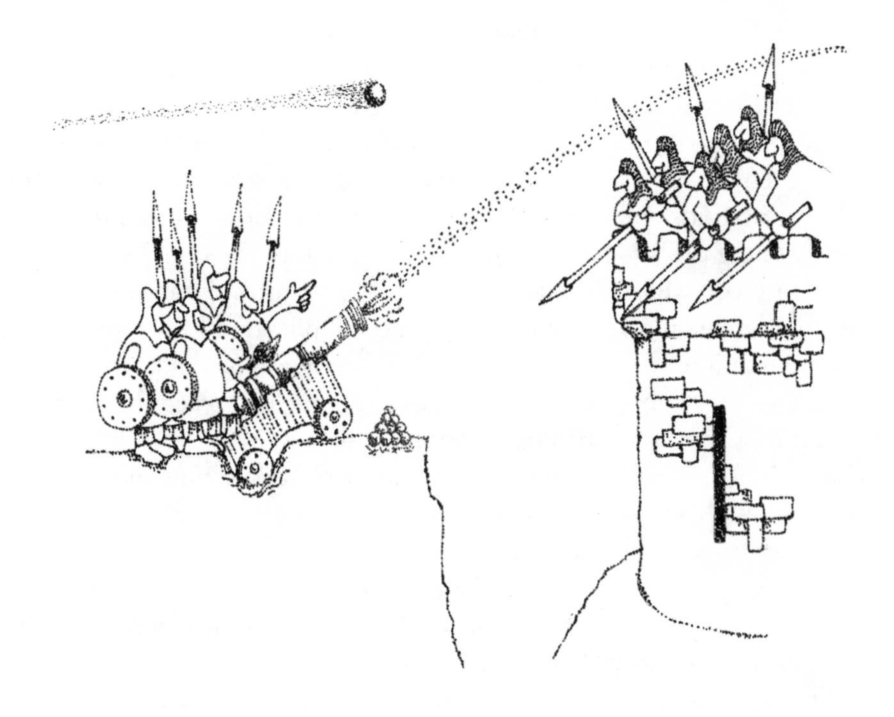

FIGURE 3: The general orders ten sacks of powder. "KA-BLAM." The ball sails into the sky, and ninety minutes later it sails overhead in orbit. The ball is continuously accelerating downward at g = 32 feet/second^2, but Earth's ground is receding at the same rate (relative to the ball) since Earth's surface is curved, and the ball is moving forward with a huge velocity. This maintains the continuous circular orbit. (Illustration by Ilse Lund.)

This "thought" experiment neglects the resistance of air but is otherwise exactly as Newton's equations would predict. Earth's gravitational pull always impels the cannon ball to "fall" toward Earth, but the starting conditions are different in each instance. A low initial velocity would cause the ball to fall more or less vertically. Higher velocities would give it the trajectory of near-Earth projectiles. The higher the initial velocity, the farther the ball would fly before landing back on the surface of Earth again. Since Earth's surface is curved, however, there comes a certain velocity such that the "fall" toward the earth is exactly matched by the curvature of Earth: at that point it is in orbit. If we had added two or three more bags of special high-power gunpowder, the ball would arc away from Earth and escape from its gravitational pull altogether. The basic equation is always the same—but the diverse starting conditions would generate the variety of different outcomes that we'd get, from the paths of asteroids and comets to those of planets, *Voyager* satellites, and fools jumping off bridges with bungee cords tied to their ankles.

In case you were wondering, Newton's amazing and universal equations also have some profound philosophical implications. If we know the initial conditions of anything—in the cannon example these are (*1*) where the cannon ball is at one time, and (*2*) in what direction and how fast it is moving (how much powder we use)—then we should be able, in principle, to predict precisely its entire future. Predicting the future? Now that really challenges Aristotelian philosophy!

For example, if we know the exact initial positions of the nine planets of the solar system (how far each is from the sun) and their exact velocities, and we are given the precise forces between them (which depend on their masses), and if we have a monstrously powerful computer, we can predict the entire system into its distant future to whatever precision we desire. The next step is an even bigger one: if we know the initial conditions of each particle in the swirling hot dust clouds composing the embryonic solar system, we should be able to predict the future formation of the planets and their moons. All things are predictable in classical physics given enough computer power and precise knowledge of initial conditions. We even have a word for it: we say that classical physics is *deterministic*. In classical physics the future can, at least in principle, be precisely determined. Remember this important fact when we get to the quantum revolution.

NASA relies on and encodes Newton's laws into its computer programs to predict the complex orbits for its planetary satellites. Students at Caltech, MIT, and other institutions apply these laws to mechanical, civil, and architectural engineering. These laws make space travel possible and allow us to design bridges, skyscrapers, automobiles, and aircraft. They have enabled modern civilization to become the thriving, multifarious complex it's become.

So what is wrong with Newton's theory? Simple! Despite three hundred years of satisfied customers, the Newtonian system fails in two domains: the domain of huge velocities (things approaching the speed of light) and the domain of the very small (the scale of the atom). What does work inside the atom is the quantum theory.

Chapter 3

LIGHT AND ITS VARIOUS CURIOSITIES

Before we leave classical physics behind, we need to spend a little time pondering and playing with light. Many important and initially baffling questions about light will pop up again in a new guise when we delve into the quantum realm. In the meantime, let us look at the origins of the theory of light within the classical framework.[1]

Light is a form of energy, and various processes can change electrical energy into light (e.g., a toaster or a lightbulb) or chemical energy into light (a candle or fire). Sunlight comes from the intense heating of the solar surface due to the energy produced by processes deep within the sun, called nuclear fusion. Radioactive particles emitted from a nuclear reactor core generate a faint blue light in surrounding water as they rip apart (ionize) atoms.

A small amount of energy injected into a chunk of any substance, such as a block of iron, will heat it. At low levels of injected energy, this may be a warmth that can be felt by your hand (even Sunday-afternoon carpenters know that a nail heats up when hammered or pulled out of a block of wood). The block of iron when heated sufficiently gives off dim radiant energy in the form of a dull reddish-colored light. As the temperature increases, orange and yellow colors are added to the red, and at still higher temperatures, green and blue join the mix. The result, if things get hot enough, is a bright white light emitted from the material, a mixture of all colors.

We can see most objects around us, not because they emit light, but because they reflect light. Except for a smooth mirror, this reflection of light is imperfect. A red object receiving white light from the sun reflects only the red part and absorbs the orange, green, violet, and so on. Various pigments are chemicals that behave differently in their absorption of light. Adding pigment to a material is a mechanism for selectively reflecting some colors, whereas the other parts of the color spectrum that aren't reflected are absorbed. White objects reflect all colors, while black objects absorb all colors—which is why the asphalt pavement of a parking lot gets so hot on a sunny day and why wearing a white shirt rather than a black one is more comfortable in the tropics. These phenomena of absorption, reflection, heating, and their relationship to various colors of light can all be measured by various scientific instruments and can be expressed as numbers.

Light is full of curiosities. I "see" you across the room, which is to say that light reflected from you beams its way to my eyes. How nice! But your friend, Edward, is looking at the piano, and the light beam from the piano crosses the you-to-me light beam with no apparent disturbance. Light beams (invisible in the absence of chalk dust or cigar smoke in the air) pass through one another easily. However, when the two beams from two flashlights, for instance, light up an object, it is twice as bright as when illumined by only one beam.

Let's look at a fish tank. With a pocket flashlight, a darkened room, and some dust obtained by slapping together some blackboard erasers or a dust mop, you will see the beam of light, reflected in the air by the chalk dust, bend as it obliquely hits the water (and you may see one confused Blue Gourami expecting his fish food). This bending of light by transparent materials such as glass and plastic is called *refraction*. Boy Scouts use a magnifying glass to focus a beam of sunlight on a tiny spot of wood to start a fire. They are taking advantage of refraction of light by the lens, as each individual ray of light is bent to converge toward a point, called the "focal point." Thus they succeed in concentrating the light energy to rapidly heat the wood and cause combustion.

A glass prism hanging in our window splits the white sunlight into its spectral constituents: Red-Orange-Yellow-Green-Blue-Indigo-Violet (ROY G. BIV). Our eyes respond to the visible colors, but we know that the energy continues beyond what is visible. Invisible on one end of the spectrum is the long-wavelength infrared light (such as is produced by an infrared heat lamp, warm toaster wires, or the embers of a dying fire) and on the other end is the short-wavelength ultraviolet light ("black light," or the intense light produced by a welding torch, which requires the welder to wear his goggles). White light is therefore composed of equal amounts of the different colors of light. We can take the colors of light and combine them to make white light. With our measuring devices we can selectively determine the intensity of each color band (the bands of colors represent the "wavelength" of the light). The amount of light for each wavelength can then be plotted in a graph. When we do this for hot, glowing objects that are emitting their own light, we find that the resulting graph forms a bell-shaped curve, peaked around a given wavelength (color) of light (this is shown in figure 13). At low temperature the peak of the curve is in the long wavelength, or red light. As we increase the temperature, the peak of the energy curve moves toward the blue part of the spectrum, but there is still enough of all the other colors present to make the object glow white. At still higher temperatures, the object glows bluish-white. On a clear night you can look at the stars and see slight color variations. The reddish stars are cooler than the white ones, which in turn are cooler than the blue stars. This represents the different stages of the evolution of stars as they burn through different nuclear fuels. And this simple result was the birth certificate of the quantum theory, about which we'll have much to say later.

HOW FAST DOES LIGHT TRAVEL?

It is not immediately intuitive that light is an entity that must travel across the space between a luminous source and your eyeball. To a child, light doesn't seem to travel at all; it simply shines. But travel it must, and Galileo was one of the first to try to pin down the velocity of light, employing two fellow assistants who worked all night on adjacent mountains covering and uncovering lanterns at exact moments of time. They counted out loud and tried to discern a delay as the distance to an observer (Galileo) was increased. You can clearly measure the speed of sound this way—witness a bolt of lightning striking the city water tower a mile away and count out the seconds until the boom of thunder arrives. The speed of sound is a measly one thousand feet per second, so it will take about five seconds for the thunder clap to travel the roughly five thousand feet in a mile, an easy time delay to count out loud. But Galileo's humble experiment to measure the speed of light failed since the speed of light is much too large to measure in this way.

In 1676, a Danish astronomer at the Paris Observatory named Ole Römer, used his telescope to measure in detail the motion of the moons of Jupiter (these are the "Jovian moons," which Galileo had discovered nearly a century earlier).[2] Römer observed eclipses of the moons by the great planet and discovered there was often a lag in the time that the moons disappeared and reappeared from behind Jupiter at the beginning and end of an eclipse. This time lag mysteriously depended on the distance of Earth from Jupiter that changed over the course of an earthly year (for example, the Jovian moon Ganymede reappeared earlier in December and later in July). Römer realized that he was observing an effect due to the finite speed of light, just like the later arrival in time of a thunder clap from a more distant bolt of lightning.

When his careful measurements of the time delays were combined with the first accurate measurement of the Earth-Jupiter distance in 1685, it yielded the first precise measurement of the speed of light, a whopping 300,000,000 meters per second (or 186,000 miles per second, compared to the speed of sound at 0.2 miles per second). Later, in 1850, two very skilled, but highly competitive, French scientists, Armand Fizeau and Jean Foucault, succeeded in making the first precise non-astronomical measurements of the speed of light on planet Earth. Then the "catch-me-if-you-can" competition of better and more precise measurements of the speed of light commenced in the scientific world and continues to this very day, with a current best value of $c = 299{,}792{,}458$ meters per second. Notice that in physics we always call the speed of light "c." So whenever you see an equation like "$E = mc^2$," you will recognize that "c" is the speed of light, and it is one of the most important pieces of the puzzle that constitutes the whole physical universe.

BUT WHAT IS LIGHT MADE OF? PARTICLES OR WAVES?

So light travels (we say "propagates") through space from one point to another with a really large speed. But we must still determine something very fundamental about light: what is the light that is propagated from your neighbor's lilac bush to your eyes? In general, what is light? Our intuition about the world is that all things are made of smaller bits, which we can call particles. So one very plausible idea is that light is actually a stream of particles emitted from the light source and collected by your eye and squeezed onto the retina, where other biochemical reactions occur that create a sensory experience called "vision" in your brain.

Particles are a good hypothesis of what light is since they can carry energy. Particles can scatter, that is, be reflected by surfaces; they can induce chemical reactions. But they must also have some kind of an internal structure that generates color. Like Galileo before him, Isaac Newton was convinced by his interpretation of all the available data of his time that light is propagated as "a shower of tiny invisible particles." These particles, after being emitted from a light source, travel at enormous speeds in straight lines until they collide with materials that cause them to be absorbed, reflected, or refracted. Remember, this was about 1700, a time that the speed of light had actually been measured, so Newton, unlike Galileo, actually knew that light propagation was not instantaneous. Theorists, of which Newton was one of the greatest, always need strong reinforcement for their theories from experiment. Newton concluded that refraction—the bending of light by glass or water—is induced by a change of velocity of light particles in the glass, water, or any other refracting material.

Why would refraction occur? Picture a Newtonian light particle heading at an angle toward a piece of glass or the surface of water. The idea was that the glass or water medium "tugs" at the particles of light as they reach the surface boundary, which subtracts some component of the motion in the direction along the surface. Subtracting that piece of velocity from the original results in a "bending" of the stream of particles—a plausible argument for explaining refraction.

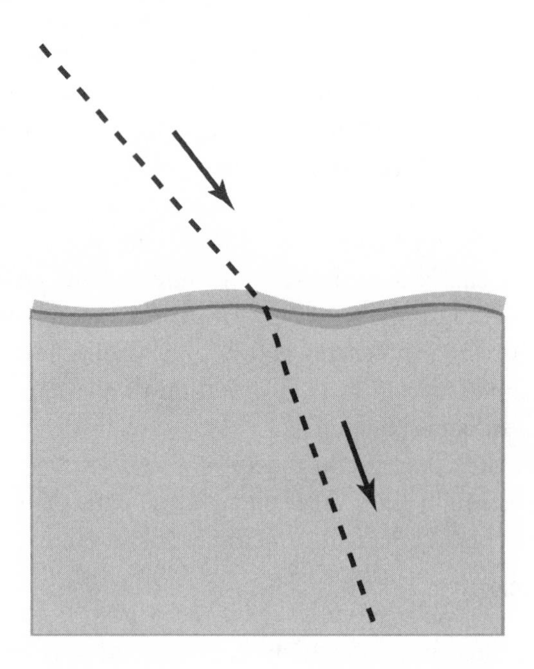

FIGURE 4: Refraction of a light beam traveling from air into water.

This wasn't the only theory at the time, however. A competing idea made an analogy with the phenomenon of sound that was known to propagate through air as a pressure disturbance that travels in waves, much like the waves that propagate along the surface of water. According to this hypothesis, perhaps the entire universe is filled with some transparent substance, and light is a wave disturbance moving through this medium. Newton's contemporary, Christian Huygens, indeed believed that light propagates as waves, behaving much like the circular waves that spread outward when you tap the surface of a still pond with your fingertip. He showed that waves would naturally also bend (refract) as they pass from space into a dense medium, provided that the speed of light is reduced in the dense refracting medium, compared to that in free space.

Actually, light does move more slowly in a refracting glass or water medium. Since no one could measure the speed of light at the time, other than in astronomical settings, this crucial aspect of the theory could not be put to the test for another 150 years. Although both theories fit the data of the time, such was the force of Newton's authority over science and scientists that his light particles, which he called "corpuscles," became the standard theory . . . until 1807, that is.

THOMAS YOUNG

In that year, a polymath English medical doctor with a passion for physics performed an unforgettable experiment. Thomas Young (1773–1829) was a wunderkind who learned to read at age two and by age six had read the Bible through twice and had started the study of Latin.[3] In boarding school he gained a reading knowledge of Latin, Greek, French, and Italian; began to study natural history, philosophy, and Newton's calculus; and learned to make telescopes and microscopes. While still in his teens, Young tackled Hebrew, Chaldean, Syriac, Samaritan, Arabic, Persian, Turkish, and Ethiopic. Between 1792 and 1799 he studied medicine at London, Edinburgh, and Göttingen. Along the way he abandoned his Quaker upbringing and reveled in music, dancing, and the theater. He bragged that he had never spent an idle day in his life. Obsessed with Egyptology, this extraordinary gentleman, scholar, and autodidact was one of the first people to translate Egyptian hieroglyphics. He persisted in compiling his Egyptian dictionary until his dying day.

Unfortunately, Young was never very successful as a doctor, perhaps because he did not inspire confidence, or did not have that *je ne sais quoi* in his bedside manner with his patients. His languishing London practice, however, gave him plenty of time to attend meetings of the Royal Society and discuss ideas with the scientific bigshots at the time. For our purposes, Thomas Young's greatest contribution was in the field of optics. He started his research in 1800, and by 1807 he had performed a series of increasingly decisive experiments supporting the wave theory of light. But before we get to his most famous experiment, we need to take a brief look at the behavior of waves in general.

Let's examine water waves, beloved by surfers and Romantic poets. Picture the waves, way out in midocean. Measuring the distance between crests will give us the *wavelength*, while the height of the crest above a calm sea surface is called the *amplitude*. Crests will measure so many feet, or meters, above "zero," and troughs an equal number below zero. The waves travel, their crests moving through the sea, at a *wave velocity* (which would be *c* for "light"). A progression from one crest down to a trough and back to a crest again is called *one cycle*. This brings us to *frequency*, the rate at which crests (or troughs) pass a given point in space. If three crests pass in a minute, then the frequency is 3 cycles/minute. The wavelength—the distance 3 between crests (let's say, thirty feet) times the frequency (let's say, 3 cycles/minute) equals the wave speed—in this case is, 90 feet per minute, or about 1 mile per hour.[4]

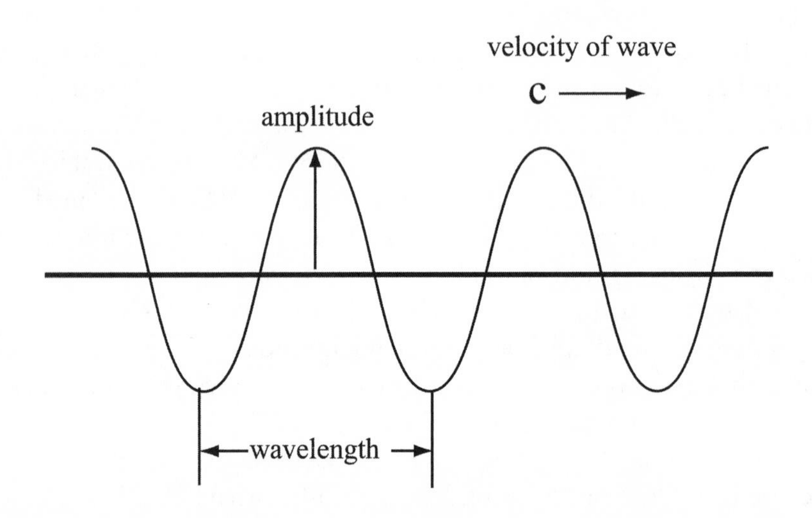

FIGURE 5 : A wave-train, or "traveling wave." The wave moves to the right at a speed of *c* and has a wavelength (length of one full cycle, from trough to trough, or crest to crest). A stationary observer watching the wave travel past would see a frequency of *c*/(wavelength) crests, or troughs, passing by per second. The amplitude is the height of a crest above zero.

The wave frequency is a familiar characteristic of sound and is very audible to human ears. Sound waves vary from a very deep basso of 30 cycles/second up to the limit of human hearing at a squeaky 17,000 cycles/second. "Concert A" is the A note on the piano keyboard immediately above middle C in the musical scale and has a frequency of 440 cycles/second. The speed of sound in air, as we have seen, is about 1100 feet/second, or 770 miles per hour. Applying simple math—the wavelength is equal to the speed of sound divided by the frequency—we can deduce that the wavelengths of concert A is (1100 feet/sec.)/(440 per sec.) = 2.5 feet. All of the audible-to-humans sound waves range from wavelengths of about (1100 feet/sec.)/(17,000 per sec.) = 0.065 feet to (1100 feet per sec.)/(30 per sec.) = 37 feet. It is the wavelength, as well as the speed of sound, that dictates how sound rattles around canyons, travels through the open air of Wrigley Field, or fills a concert hall.

The world is full of different kinds of waves: water waves, sound waves, waves on ropes and in springs, and seismic waves that shake the earth under us. All these waves can be described in terms of classical (nonquantum) physics. The amplitudes of the waves in each case represent different quantities—the height of water, the pressure (in pounds per square inch) of sound waves, the displacement of the rope or compression of the spring, and so on. All these involve a disturbance, or the deviation from the normal of an otherwise undisturbed medium. The disturbance, like a pluck of a long string, propagates away in the shape of a wave. In the realm of classical physics, it is the *amplitude* of the wave that determines the energy carried by this disturbance.

Imagine a fisherman sitting in his boat on a lake. He tosses a fish line in the water attached to a "bobber." The bobber allows a fixed amount of line to drop into the water without touching bottom and provides a visual signal to the fisherman if a fish is on the line. The bobber bobs only up or down as the waves move by. A regular cyclic change in the position of something, like the bobber, that continually repeats itself—starting from zero, rising to a peak, then sinking through zero down to a trough and back up to zero—is known as a *harmonic* wave, also known as a *sine wave*. We'll just call it a wave.

DIFFRACTION

Now we'll add another phenomenon and another, and very important, word to our wave vocabulary: *diffraction*.

Consider a harbor protected by a sea wall with a narrow opening for the passage of ships. Ocean waves with long parallel crests come in from a great distance and hit the sea wall, where they crash and break. The waves that hit the narrow opening (the opening is "narrow" compared to the wavelength of the waves), however, pass through and spread out into the harbor in all directions. It is as if the narrow opening is a source for waves, as if someone was tapping his finger in the middle of a pond, making the waves emanate outward equally in all directions. This phenomenon of a wave spreading out in all directions from a small aperture is known as *diffraction*. Sound waves do it, too, which is why we have no trouble hearing around corners. Careful measurements show that the amount of wave spreading depends on the wavelength and the size of the opening. The longer the wavelength and the smaller the hole, the more

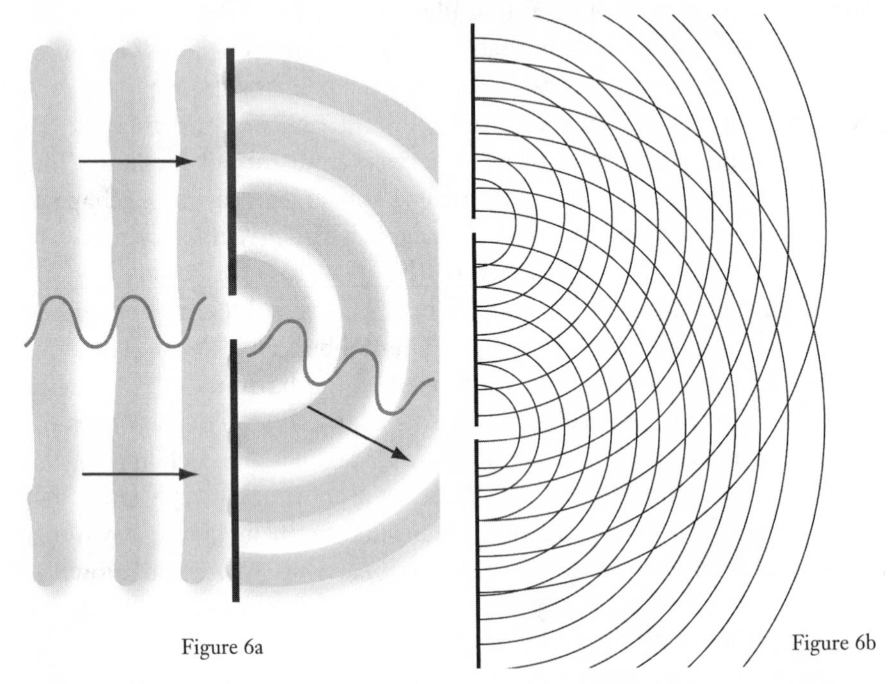

FIGURES 6a and 6b: The diffraction of a train of waves entering a narrow harbor entrance (a). The same thing happens to light, sound, or any other wave. The diffraction of sound is what allows us to hear sounds around corners. Single-slit diffraction of light is due to the finite width of the slit opening. By combining waves from two slits we get the diffraction pattern seen in (b). When light waves pass through two slits and impinge on a screen, we get the alternating bright-dark bands of the diffraction pattern observed by Thomas Young.

spreading, while as the opening becomes larger than the wavelength, the waves pass through more or less maintaining their original direction.

You can test this for yourself by doing various experiments in the bathtub with waves. Try to duplicate the long wavelength diffraction as water waves pass through a narrow harbor entrance. Or, if you're a good observer and have good eyes, you can see diffraction by looking at the light from a lamppost at night. If you squint to reduce the hole through which the light enters your eye, you'll see a flaring out of light streaks—an example of diffraction.

One reason the wave theory of light was so late in coming was that no one had ever convincingly seen light beams diffract, changing direction as they passed through a tiny hole. Ergo, everyone thought, light was not a wave. But Young's earliest arguments insisted that the wavelengths of the light waves were very tiny (say, one hundred thousandth of an inch), so the diffraction of light in passing through holes was extremely slight and would have escaped observation.

We should mention one final aspect of waves: the phenomenon of *interference*. Waves can be added (or subtracted). This is what happens when they occupy the same place in space. Two things can happen: troughs can be in the same place as crests and cancel each other, or crests (or troughs) can pile up on top of one another to make a humongous wave. In fact, such waves where many troughs have just randomly piled on top of one another can occur at sea and are known as rogue waves, and they pose a tremendous hazard to ships.[5]

Waves that have nearly the same wavelength (hence nearly the same frequency) are most efficient at adding up or canceling each other out over larger regions of space. We say that two waves whose crests arrive simultaneously at the same point are "in phase" and result in a giant wave with twice the amplitude of each alone. Waves can also arrive "out of phase" so that they cancel each other out—generating a wave of zero (flat) amplitude. There is a continuum of possibilities between "in phase" and "out of phase," and therefore a continuum of possible amplitudes, at the point where the two waves arrive. Since two waves can *interfere* with each other in this fashion, we call this phenomenon *interference*. The adding together of crests or troughs to form a larger amplitude is called *constructive interference*, while the canceling of troughs against crests is called *destructive interference*.[6] We are now ready to appreciate Young's double-slit experiment.

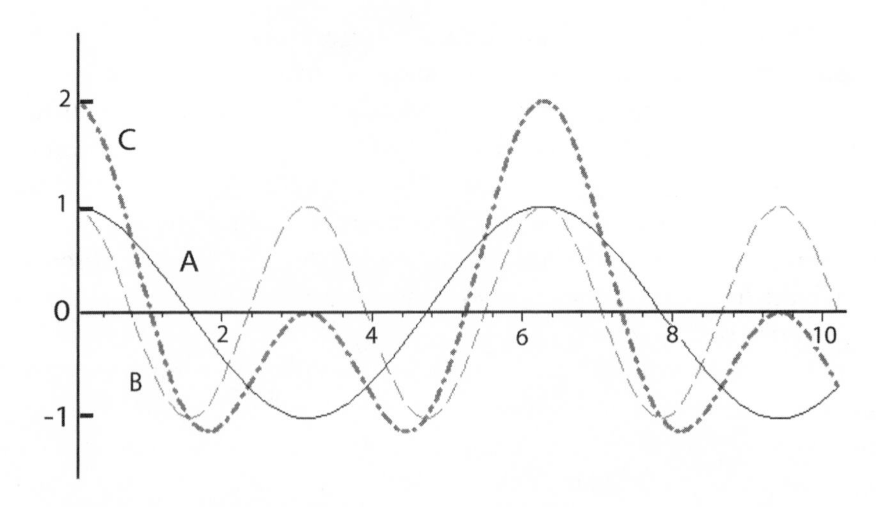

FIGURE 7: We see how two waves can combine and interfere. If we mathematically add wave (A, solid line representing cos(x)) to wave (B, dashed line representing cos(2x)) we get the result of (C, dot dashed). Note that (C) has alternating higher peaks and lower peaks. By adding more waves together we can generate any pattern for the sum that we want (Fourier analysis).

THE ECSTASY AND THE AGONY OF YOUNG'S DOUBLE-SLIT DIFFRACTION EXPERIMENT

This was the first of many experiments illustrating the wave property of interference, and it produced data that were totally inconsistent with Newton's concept of light as "corpuscles," or a stream-of-particles. Incidentally, Young realized it was risky to challenge the grand icon of physics, so he cleverly prefixed his arguments with selections from Newton's own writings that expressed some doubts on the subject of the wave-versus-particle model.

To reproduce Young's experiment demonstrating the wave property of light, we could use an inexpensive laser pointer as a stationary projector. We would direct its beam to a screen, perhaps aluminum foil with two vertical narrow slits that are close together. The slits would be very fine, for example, cut with a razor blade in aluminum foil or etched on a smoked glass. The slits would be parallel to each other about a millimeter apart (the closer the better). The light shining through the two slits would be allowed to propagate to a second screen, a distance of, say, ten to fifteen feet away. In a darkened room we would observe the light falling on the distant screen. We would observe the light falling on the second screen. There we would see a pattern that consists of a series of alternating light and dark bands of illumination (figure 8). The bands are parallel to the slits on the first screen. In other words, there are places on the screen that are quite bright where crests of light have reinforced one another, and other places where crests have canceled troughs and no light arrives. This is a pattern called *interference fringes*.[7]

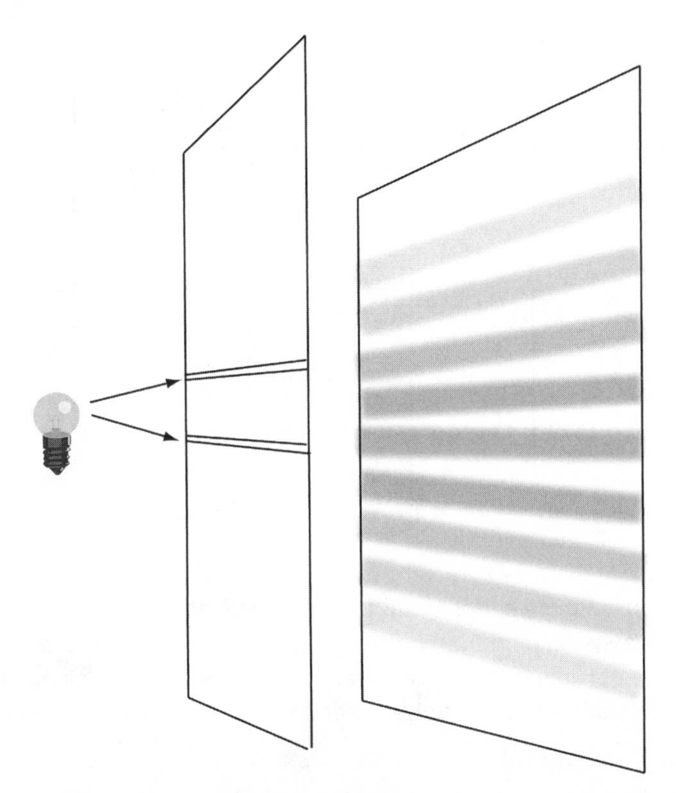

FIGURE 8: The interference pattern of light waves passing through two slits, as in figure 6b, and projected onto a distant screen is illustrated. The observation of this phenomenon is what enabled Thomas Young to prove that light was a wave phenomenon.

If we closed one of the two slits, we would get something totally different: we would get only a swath of light opposite the open slit and fading into darkness on both sides (see figure 9). The dramatic interference effect is seen only when both slits are open (figure 8) and the waves from the two slits can overlap on the distant screen, producing the bright and dark interference fringes.[8]

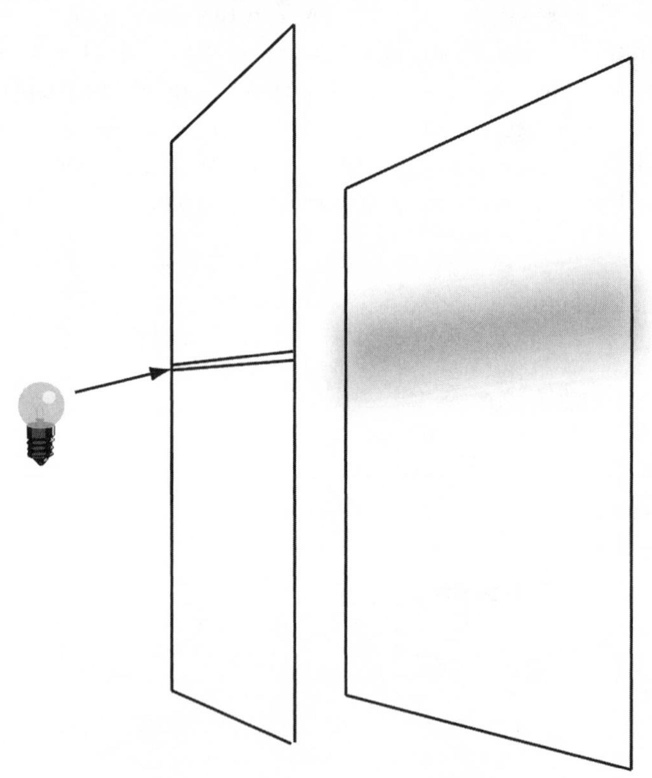

FIGURE 9: In Young's experiment, the observed interference is due to the addition of waves coming from both slits as in figure 8. If we close off one slit, we no longer observe the interference effect (there is a "single-slit" interference, but it isn't observable when the slit opening is made to be extremely narrow).

What does it all mean? Imagine that we have a small detector of light at some point on the second screen, labeled **P**. Light reaches **P** from both slits. Since light is a wave, the light reaching the slits from the source may be at a crest, a trough, or some intermediate phase. The light disturbances, as they emerge from each slit, have the same phase. If **P** is equidistant from the two slits, the two waves arrive in phase, producing a bright interference band. Now move **P** along the screen. At some position the difference in the distance from Slit **A** and Slit **B** to **P** will be such that the light waves interfere destructively, in other words, the waves will be precisely out of phase and will cancel each other out, and we will get a dark band on the screen (see figure 10).

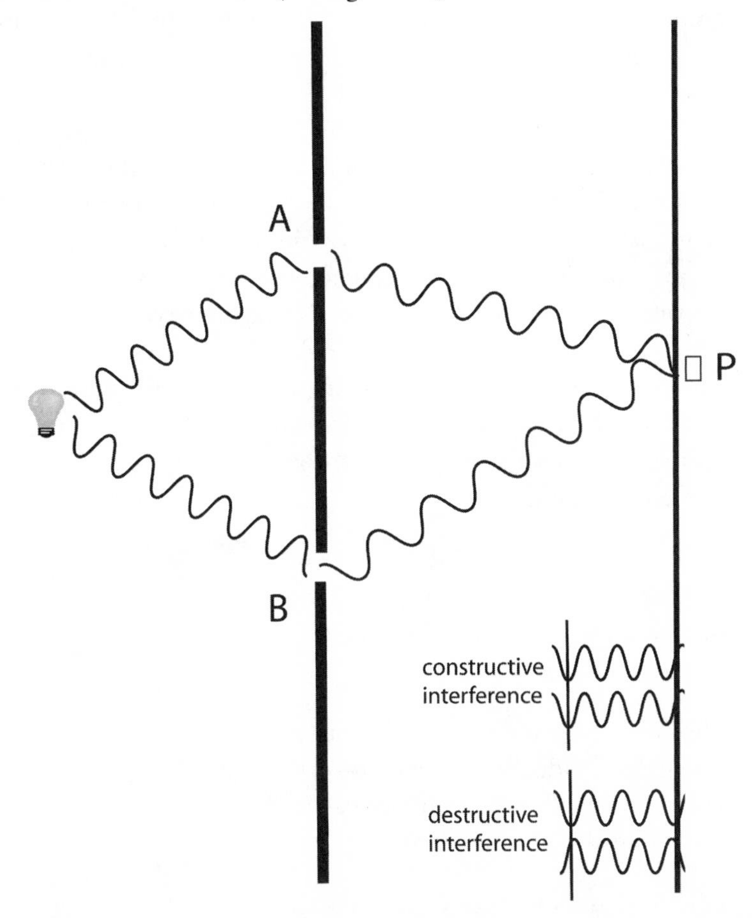

FIGURE 10: Detail of the origin of two-slit interference. At some point, **P**, the two waves interfere "constructively" on the detector screen producing a bright band. At another point, **P**, they interfere destructively, producing a dark band.

The dark and light fringes on the screen are due to the difference of the phase of two waves. When the difference of distances from each slit to **P** corresponds to one cycle (or one wavelength), we get a bright fringe. When the difference of distances from each slit to **P** corresponds to one half of one cycle (or one half wavelength), we get a dark fringe.

Young's brilliant and simply stated argument was that, under certain circumstances, adding two beams of light can result in darkness—destructive interference where a wave crest adds to a trough. This is a classic feature of an interference pattern, and it screams "wave phenomenon." If you want to observe it for yourself, simply notice how an oil slick or gasoline spill on a puddle produces brightly colored striations. The reason is interference. Light is reflected from the top surface of the thin oil film as well as from the underlying surface, and their phases differ as one ray passes though the film twice. When combined in your eye, the two beams interfere and, if things are just right, they can cancel each other. But in this case we're dealing with white sunlight (containing all wavelengths), and the waves can cancel for only one wavelength—say, red. So red cancels, and your eye sees what is left when you subtract red from white: a blue color. At slightly different thicknesses of oil other wavelengths cancel, giving rise to the brilliant colors you see. Or wait for a rainy day and find a rainbow. You'll be observing primal nature and interference of light passing through tiny droplets of water in the air at different angles.

On this long storm the Rainbow rose—
On this late Morn—the Sun—
The clouds—like listless Elephants—
Horizons—straggled down—

The Birds rose smiling, in their nests—
The gales—indeed—were done—
Alas, how heedless were the eyes—
On whom the summer shone!

The quiet nonchalance of death—
No Daybreak—can bestir—
The slow—Archangel's syllables
Must awaken her!

Emily Dickinson,
"On This Long Storm the Rainbow Rose"[9]

It was difficult to imagine how Newtonian particles could cancel one another here and add up over there, forming an interference pattern. If we pile apples on top of apples in our bushel basket, we always get more apples—not fewer!

YOUNG'S CONCLUSION: LIGHT IS A WAVE

About a decade later, a French physicist, Augustin Fresnel, confirmed and extended Young's results, and the idea of light as waves was established. The subject of wave optics expanded and led to the design of many refined optical instruments, from microscopes to telescopes, based on light waves.

The wave theory seems to account for all the diverse phenomena we might encounter: reflection, absorption, refraction, diffraction, and especially interference. According to the wave theory, toward the end of the nineteenth century, light is emitted as the vibrations of atoms. At the time, this was dimly understood, but it was known that these vibrations were necessarily very fast, measuring a thousand, trillion (10^{15}) cycles per second, which corresponded to the frequencies of light waves. Remember, the frequency is the speed of light divided by its wavelength. The frequency one gets by dividing a tiny wavelength into a huge velocity is achievable only by these rapid vibrations at "atomic"-size scales. They knew that colors are the physiological effect of the wavelength of light absorbed in the retina. Multiply the wavelength by the vibratory frequency, and you get the velocity, which, in a vacuum, is the same for all wavelengths (all colors), the speed of light, our ubiquitous "*c*." We should note that the velocity of light in the vacuum is the same for all sources, from candles and glowing metals to the light of the sun. But when passing through most materials, like glass or water, light slows down. Different wavelengths (colors) travel at slightly different speeds in materials. This permits us the luxury of separating white light into its constituent colors, as happens with a prism.

Although most of the known phenomena of light by the twentieth century fell happily into place when described by the wave model of light, there were also some discrepancies.

UNANSWERED QUESTIONS

A host of residual questions were still not satisfactorily answered at this stage of the development of the theory of light: What is the mechanism by which light is generated? What is the mechanism whereby light waves are absorbed, and why, in the case of colored objects, are only certain bands of colors (wavelengths) absorbed? What secret operations occur in the retina, or on a photographic plate, enabling us to say, "I see"? These questions all deal with the interactions of light and matter. Moreover, how does light propagate through the vacuum of space between the sun and Earth—given that our experience with sound and water waves tells us there must be a responsive medium to propagate a wavelike disturbance? Must there be some kind of ghostly, transparent, weightless medium filling the vast stretches of space? Nineteenth-century physicists gave it a name: the *ether*.

And ponder still another mystery, this one concerning the sun. This super light-wave generator is a powerful source of both visible and invisible light, the latter comprising the longer-than-visible wavelengths, starting with infrared, and the shorter-than-visible wavelengths, starting with ultraviolet. The atmosphere—primarily the ozone of the upper stratosphere—filters out much of the ultraviolet and all of the even shorter wavelengths, such as x-rays. Now suppose we invent a device that, at the flick of a dial, can select out a small band of wavelengths, absorb the light, and measure its energy.

We do have such a device (usually there's one in every well-equipped high school science lab): it is called a *spectrometer*. A spectrometer bends red light the most, violet the least, and it fans out all the colors away from the direction of the original beam of white light. Newton's glass prism was an early, primitive spectrometer. We add a viewing-scope, mounted on a nicely engraved scale, that measures the angle that the viewing-scope makes with the original white light beam. Since the color (wavelength) of light determines the degree to which it is bent, we can easily translate angles into wavelengths.

Now let's move the viewing-scope to the place where the deepest red fades to black, in other words, where there's no visible light. The scale reads 7500 Å, where "Å" stands for *angstrom* units, after the Swedish physicist Anders Jonas Ångström, who helped develop the field of spectroscopy. One "Å" measures a hundredth of a millionth of a centimeter, 10^{-8} cm. So we have determined that the deepest red light is a wave measuring 7500 Å, or a few hundred thousandths of an inch, from crest to crest. This is one end of the visible spectrum; if wavelengths get any longer, we need detectors that are sensitive to infrared and to the longer wavelengths. Now rotate the telescope to the short-wavelength end of the visible scale—to the dim violet, say, about 3500 Å. Below 3500 Å we need something other than an eyeball to detect the ultraviolet wavelengths.

So far, this is all just a fine-tuning of Newton's discovery of the spectrum of colors in white light. But, using such an instrument in 1802, an English chemist, William Wollaston, found that when looking at the sun, the color spectrum that smoothly flows from deep red to deep violet is overlaid by many very fine dark lines. What were these dark lines?

Enter Joseph Fraunhofer (1787–1826), a highly skilled though unschooled Bavarian lens maker and optical scientist.[10] The eleventh and last child of an impoverished glazer, Fraunhofer received only a rudimentary elementary education before being sentenced to Dickensian toil in his father's workshop. After his father's death, the sickly youth entered a degrading apprenticeship to a Munich mirror-maker and glass cutter. In 1806, he landed a position in the optical shop of a Munich scientific instrument company, where, under the tutelage of a trained astronomer and an optics expert, he mastered practical optics and acquired an expertise in mathematics and optical science. Discontented with the poor quality of the optical glass then available, the perfectionist Fraunhofer managed to negotiate a contract that permitted him to penetrate the closely guarded professional secrets of a great Swiss glassmaking firm, which had recently relocated to Bavaria. This collaboration resulted in superior lenses and, more important from our perspective, a theoretical breakthrough that would put Fraunhofer's name in the scientific history books.

In his quest to make lenses that approached the optical ideal, he hit on using the spectrometer to measure the light-bending power of different glasses. Turning his fine instrument to the sun, he was amazed at the proliferation of the fine dark lines mentioned by Wollaston, and, by 1815, he had meticulously recorded the precise wavelengths of many of the nearly six hundred lines he counted. He labeled the most prominent lines with the bold capital letters **A**, **B**, **C**, through **I**; **A** being a dark line in the red, and **I** in the far reaches of violet. What was going on? Fraunhofer was aware that a metal or a salt placed in a hot flame gave off a characteristic color. When studied with the spectrometer, it revealed a series of fine bright lines with wavelengths in the region of the prominent color.

Intriguingly, he noted, the pattern of bright lines from the salts precisely matched the pattern of dark lines in the solar spectrum. Table salt, for instance, emitted several bright yellow lines in the **D** region of Fraunhofer's map. In due course, a plausible explanation arrived. Recall that a discrete, well-defined wavelength corresponds to a unique, precisely defined frequency. Clearly, something in matter, presumably something at the atomic level, liked to vibrate at certain definite frequencies. Atoms (not confirmed to exist nor understood at all at the time of Fraunhofer) evidently have visible fingerprints!

FINGERPRINTS OF THE ATOM

Consider this familiar mechanical phenomenon in music: A tuning fork of concert A above middle C rings out at precisely 440 cycles per second. In the tiny realm of the atom, the frequencies would be vastly higher, but in Fraunhofer's time it was possible to imagine the mysterious atom chock-full of teensy, tiny, frantic tuning forks, each with its own frequency but now emitting light of a wavelength corresponding uniquely to that frequency.

But what about the dark lines, you may ask? Well, if atoms of sodium, excited by a hot flame, vibrate at frequencies corresponding to emitted light at wavelengths of 5962 Å and 5911 Å (both in the yellow range), then the same atomic structure would preferentially absorb light at those same wavelengths. The white-hot surface of the sun emits light of all wavelengths, but this light then passes through the relatively cooler gases of the sun's outer atmosphere (the "corona"). Here the atoms might absorb just those same wavelengths they like to emit. This absorption accounts for the curious dark lines observed by Fraunhofer. Bit by bit, spectroscopy in the post-Fraunhofer era revealed that each element, when heated, produced a characteristic set of "spectral lines," some prominent (like the brilliant red lines of neon gas that are so familiar in neon signs), and some faint (like the dimmer blue of mercury vapor lamps). These lines were the fingerprints of the chemical elements, the first clues to the tiny "tuning forks" within atoms, or whatever the mysterious structures vibrating inside the atom must be.

Since the spectral lines were very fine, the spectrometer's scale could give very precise readings, for example, 6503.2 Å (in the deep red) or 6122.7 Å (in the bright red). By the late nineteenth century, thick tomes listing the spectra of the chemical elements became available, and skilled spectroscopists could identify unfamiliar compounds and detect miniscule chemical contaminations. Yet no one had any clear idea of what produced these dramatic messages from the barely understood atom.

Spectroscopy's second major achievement was more philosophical. In the sun's signature of dark lines scientists could read its chemical composition, and lo and behold, they found hydrogen, helium, lithium, and the other elements of matter that compose our planet Earth. Since then, we have analyzed the light from stars in extremely distant galaxies, always finding our own familiar elements: hydrogen, helium, and so on. The universe everywhere has the same composition and the same laws of nature, all of which hints at a common origin in some incomprehensible act of physical creation.

Meanwhile, seventeenth-through nineteenth-century scientists were also baffled by yet another problem, namely, how are forces, such as gravity, transmitted over great distances? When a horse is harnessed to a wagon, we see that the force exerted by the horse to pull the wagon is visibly transmitted through the harness. But how does Earth feel the presence of the sun, ninety-three million miles away? How does a magnet tug on a nail way over there? There is no "visible" connection here—so it is dubbed mysteriously "action at a distance." The force of gravity was postulated by Newton, and it is an action at a distance. But what is the harness that connects the sun to Earth, producing the gravitational force? After struggling mightily with the question of "action at a distance," even the great Newton had to shrug his shoulders and leave it to future physicists to solve.

MAXWELL AND FARADAY: THE LAIRD AND THE BOOKBINDER

It was Michael Faraday (1791–1867) who first illuminated the action-at-a-distance mystery with his hypothesis of the *electromagnetic field*.[11] He was a poor man's son who got a job as a bookbinder and educated himself by reading the books he bound. Fortunately, one of those books hooked him on science. With a great intuitive leap (he was very weak at math), he decided that what an electrically charged particle does is to create an entity around itself called an *electric field*. The field is a kind of pull throughout space that acts on any other nearby electric charge (although that action diminishes with the distance from the source charge). Or, in the case of a magnetic field, we envision the space around a magnet as being "strained," filled with a magnetic field and thereby able to "notify" a distant speck of iron filing of the presence of the field, producing a magnetic force.

Now, you might well object that this "field" concept is just a fancy way of saying that electric charge **a** exerts a force on electric charge **b**. The existence of fields was the subject of much thought, debate, and philosophical rumination. The field concept lent itself elegantly to simple and compelling mathematics, and by the end of the nineteenth century, people believed in electric fields (created by charges), magnetic fields (created by magnets or by electrical currents, i.e., charges in motion), and gravitational fields (created by objects with gravitational mass). Fields provide a physical visualization of forces exerted by objects at a distance, and they explain how energy can be transferred from a charge, into a field, back to a charge again. The fields could explain action-at-a-distance. Although they were invisible, fields could be measured. A little compass needle, for example, responds to a magnetic field. A little "test" electric charge would feel a force exerted by the electric field of a distant electric charge. And, the fields themselves contained energy and momentum.

Faraday's experiments of the 1820s uncovered a deep connection between electric and magnetic fields. It had already been discovered that passing an electric current through a loop of copper wire could generate a magnetic field. Faraday asked the question in reverse: Can a magnetic field produce an electric field? The answer was astonishing: A magnetic field, when it varies in time, indeed generates an electric field. Faraday discovered this by running a large electrical current through a tightly wound coil to create a strong magnetic field. Then, using a charged wire as a probe, he tested for an electric field near the electromagnet. Zilch—no electric field was seen. He then reduced his magnetic field to zero. As he did so, he noticed that the current meter jumped briefly before settling back to zero. This revealed the presence of an electric field that appeared during the time the magnetic field was decreasing to zero. When Faraday "turned on" the magnet once more, the probe registered the existence of an electric field during the time the magnetic field was increasing. "By Jove!" Faraday exclaimed.

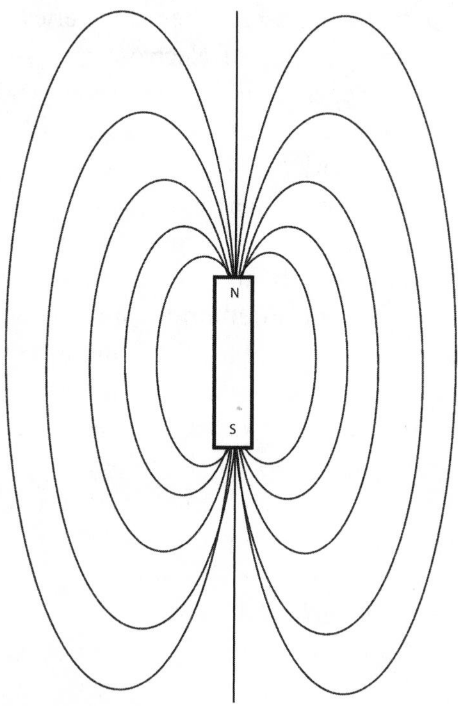

FIGURE 11: The familiar "dipole" field of a magnet. The field lines denote the direction of the field in space; the density of lines denotes the "intensity" or strength of the field. By putting the magnet under a piece of paper and placing iron filings on the paper, which will align in the direction of the field line, the magnetic field can be observed. Most physicists no longer debate the reality of "fields." They simply assume their existence and work with them.

We get an electric field produced in space when a magnetic field is present and changing in time. This wonderful discovery—called the law of induction—soon led to electric motors, electric generators, and nothing less than a modern electric society. It provides a way to convert mechanical energy directly to electrical energy. For example, a waterfall can drive a wheel to turn. The wheel is attached to a magnet and creates a changing magnetic field in coils that are positioned near the wheel. The changing magnetic field *induces* electric fields that generate electrical currents. This, in turn, permits us to send electrical energy to nearby Las Vegas or Buenos Aires. And, when we run this in reverse, a source of electrical energy can be used to spin a wheel, as in a diesel-electric train or an electric automobile, all thanks to Faraday.

Faraday's law of induction and his other discoveries laid the groundwork for a complete understanding of electromagnetism at the level of classical physics. These concepts had been expressed in assorted but largely disjointed mathematical terms. A couple of decades later, a patrician Scottish physicist named James Clerk Maxwell (1831–1879) set out to take the experimental laws of electricity and magnetism and "set them to music"—that is, to capture the complex relationships among currents, magnetic fields, and electric fields in a unified and elegant mathematical structure.[12]

The scion of a prominent Edinburgh family, Maxwell had been trained in the law, as was the custom for the Edinburgh upper crust, but his mind was seduced by technical and scientific matters. Maxwell was only fourteen years old when the *Proceedings of the Royal Society of Edinburgh* published his first paper on a methodology for drawing a perfect oval, which was similar to the string method of drawing an ellipse—a subject Descartes had also investigated in connection with the refraction of light. Maxwell showed that all the colors we see can be produced by mixtures of three spectral stimuli. He resurrected Young's theory that there are three receptors in the retina for color and proved that color blindness is due to one or more defective receptors. He demonstrated the existence of the so-called Maxwell spot in a region of the visible spectrum, which his wife found that she could not see, having almost no yellow pigment in her retina. As Maxwell wrote to a colleague, "All have it (the spot), except Colonel Strange, F.R.S, my late father-in-law and my wife."

In 1865, Maxwell laid the finishing touches to his famous "Treatise on Electricity and Magnetism" (at this same time completing a major addition to his house). In 1871, he was appointed the first professor of experimental physics at Cambridge, where he designed and developed the celebrated Cavendish Laboratory. Maxwell's lifelong desire was to understand the nature of electricity: Was it a fluid moving through wires or was it a disturbance or "strain" in a medium that must include empty space? To aid in comprehending this scenario, it was necessary to suppose that all space was pervaded by an "ether," an insubstantial medium capable of being disturbed by magnetic and electric fields.

While assembling the mathematical equations describing all the experimental laws, Maxwell made his great discovery. First, he noticed that there was (as yet) no symmetrical analogue of Faraday's law of induction: If changing magnetic fields produce electric fields, he mused, shouldn't changing electric fields produce magnetic fields? Daring to go beyond the experimental data, Maxwell discovered that, indeed, the mathematics demanded this symmetry—and his equations took on a life of their own. Waning magnetic fields created waxing electric fields, which, in turn, created magnetic fields—all propagating as a dancing wave of entwined electric and magnetic fields.

And then came a grand surprise. Plugging in the correct constants and analyzing his equations, Maxwell found that his electromagnetic fields could escape from the proximity of wires and magnets that produced them and would then move through space with a tremendous speed—one that was exactly equal to the 186,000 miles per second (300,000 kilometers per second) velocity that M. Fizeau had measured for light. Was this a coincidence? In physics, speeds like 186,000 miles per second don't just grow on trees. Maxwell arrived at his dramatic conclusion: Light is an electromagnetic disturbance, a tightly coupled mixture of pure oscillating electric and magnetic fields, propagating through space at a velocity of c = 300,000 kilometers per second.

Faraday's intuition was realized in Maxwell's theory: the fields carried energy and momentum—they were not just mathematical symbols but real physical entities. Scientists could now understand what precisely is "waving" when light waves travel across space. And they had to conclude that light affects human retinas, photographic film, and green leaves, by means of electrical and magnetic forces acting on some kind of electrically charged material that was packed inside the atoms. That atoms were repositories of electric charges was an idea that was "in the wind" among physicists. But in what arrangement?

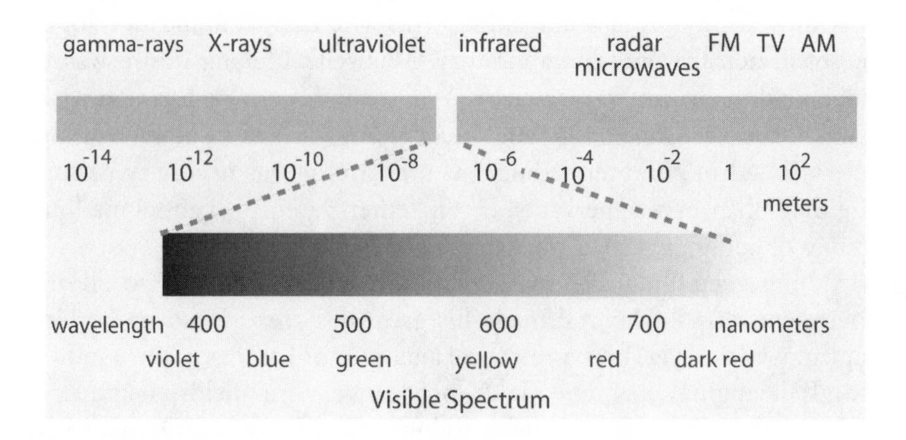

FIGURE 12: The spectrum of light. Visible light occupies a very narrow band of wavelengths, from longer (deep red) $0.00007 = 7\times10^{-5}$ centimeters (or 700 nanometers, or 7000 Å) to shorter (violet) $0.00004 = 4\times10^{-5}$ centimeters (400 nm, or 4000 Å). The energy of a photon decreases for longer wavelengths, as one descends into the infrared, microwave, and TV to AM radio wavelengths of hundreds of meters. Likewise, the energy of a photon increases for shorter wavelengths, as we go up to x-rays and very energetic gamma rays.

Between 1865 and 1880 the electromagnetic hypothesis of light was put to the test by the German physicist Heinrich Hertz, who experimentally generated electric and magnetic waves and demonstrated that they obeyed the laws of reflection, refraction, diffraction, and interference. It was a grand and spectacular success! Maxwell's equations checked out completely. Maxwell had organized his opus into four compact but highly symbolic, equations, neatly expressed by Hertz in the new mathematics of vector calculus. In the wake of Hertz, later Guglielmo Marconi, and then two world wars, Maxwell's equations would spawn the next wave of technology, bringing us Rush Limbaugh, Keith Olbermann, Oprah Winfrey, and microwaveable dinners. Visible light differs only in wavelength from the new, enormously useful radio waves, FM waves, and microwaves on one side of the visible band, and ultraviolet, x-rays, and gamma rays on the other (see figure 12).

What an incredibly spectacular scientific synthesis we now had! It seemed that all the clues fit together to account for the behavior of light and so much more. Put energy into atoms, and the little electric charges were somehow set into vibration. Maxwell's theory showed that vibrating charges radiate electromagnetic energy, including visible light. Faraday, Maxwell, and others had succeeded in explaining the universe in classical terms, with light propagating through space as waves of electricity and magnetism. Everything seemed to work fine. Nature was smooth and continuous—not corpuscular and particulate. Together with Newton's classical mechanics, Maxwell's electromagnetism gave scientists a powerful set of intellectual tools for the next great enterprise: understanding the electric charges and forces composing the chemist's atom.

And just as this massive enterprise was gathering steam, and a deep new understanding about light was setting like freshly poured concrete, the unthinkable happened: the data began indicating, with devastating clarity, that light was—a stream of particles!

The quantum spook had been sighted.

Chapter 4

REBELS STORM THE OFFICE

All the canonized laws of Galileo and Newton held sway for some three hundred years. They encompassed the beauty and rational stability of classical physics, including the golden age of the orderly laws of motion, the universal law of gravitation that governed apples and asteroids, the wonderful symmetries that underlie the theories of electricity and magnetism, and the crowning insight that light is a wave composed of electric and magnetic fields. We have just witnessed the even-keeled events leading up to the last century of the second millennium—1900—when, suddenly, things began to get a bit strange. Now we will encounter some very peculiar and eerie occurrences. We'll start with a familiar object that you face every morning before running off to work—your toaster.

We'll ask you to plug in the toaster, turn it on, and observe its internal heater wires, acquiring a warm red glow as they prepare to turn your pale English muffin to a delicious golden-brown. There is a technical term for the light you are observing: that red glow given off by the toaster wires is called *blackbody radiation.*

Blackbody radiation was, in 1900, quite the hot topic (no pun intended) in physics. That's rather remarkable given that blacksmiths and metalworkers and cooks for the previous hundred thousand years of humanity had observed this dull reddish glowing radiation emitted by hot things, such as the coals of the camp fire in front of the cave. It was only when a very smart physicist at the end of the nineteenth century sat down with Maxwell's equations in hand to try to calculate the redness of the warm glow that emanated from a blackbody radiator that he found something was quite amiss. The peculiar and detailed data about the exact properties of the light that was emitted by the wires of the toaster, and other such humble objects, ultimately shattered classical physics irreparably. Asking a subtle question about this everyday phenomenon, "Why is the glow of a campfire or toaster heating coils red?" ripped open the door to the quantum world.

WHAT IS A BLACKBODY AND WHY SHOULD WE CARE?

All objects *radiate* energy and also *absorb* energy from their surroundings. By "object" we mean something big or "macroscopic," something that is composed of, perhaps, many billions of atoms. The higher the temperature of an object, the greater the amount of energy it radiates.

Hot objects, and all of their parts (which we can think of as individual subobjects), come into a balance eventually between the amount of energy they are radiating outward and the amount they are absorbing inward. For example, place an egg from the fridge in a sauce pan full of hot water, and the cool egg will heat up, absorbing energy from the water, and the water will slightly cool down, yielding energy to the egg. Place a hot egg into cool water, and the water will heat up as the egg cools down. After a while the egg and water will have the same temperature. This is an easy experiment to do, and it is the basic example of how hot objects behave. This ultimate balance, between the temperature of the egg and water, is called *thermal equilibrium.* So, too, a particularly hot spot within an object will cool down to its surroundings, while a particularly cool spot will warm up. When we reach thermal equilibrium, all parts of the object will have the same temperature. At that point, equivalent parts are radiating and absorbing energy between one another at the same rates.

If you lie on the beach on a hot, sunny day, you will be radiating and absorbing electromagnetic radiation. There is energy radiated *to you* by that quintessential radiator of radiators—the sun. Meanwhile, your body adjusts and throws off energy to maintain the correct temperature.[1] If you are physiologically normal, at a body temperature of 98.6 degrees Fahrenheit, you continuously radiate about 100 watts of energy into your surroundings. Your body maintains a thermal equilibrium among all your parts—your liver, your brain, your toes—which is needed to sustain the chemistry of life. If the outside world is very cold, then your body needs to produce more energy or retain the energy it has produced in order to maintain this internal temperature against the loss of radiated energy to the outside world. Blood flow, which carries thermal energy to the surface, is reduced to keep our insides warm while our fingers and noses get cold. On the other hand, if it's hot outside, the body must throw off more radiation to keep cool. Warm perspiration evaporates on our skin, absorbing additional thermal energy out of the skin and providing a kind of air-conditioning effect, while transferring the energy to the outside atmosphere. A crowded room full of people gets warm: if you're stuck in a dull meeting with thirty people, the full three kilowatts of energy radiated into the room by all those people can rapidly warm the place up. But you might need to huddle with those people if your stuck in the Antarctic without a fireplace, like a wintering flock of Emperor penguins trying to protect their fragile eggs through the long winter.

People, penguins, and even toasters are complex systems. Their energy is produced internally. For people this happens by the burning of food or stored fat; for toasters, by the collisions of electrons of the electrical current with the heavy atoms in the toaster wire. The electromagnetic radiation that people and toasters radiate travels out into the external environment from their surfaces, such as human skin and the surfaces of the wires in the toaster. This radiation generally bears the imprint of definite colors of particular "atomic transitions," that is, specific chemical effects. Fireworks, for example, are definitely hot when they explode, but the chemical compounds, strontium chloride and barium chloride (and many others),[2] used in fireworks produce strong red and green colors as they oxidize, which create spectacular displays.

These are all fascinating effects, but there is a generic pattern of the electromagnetic radiation that *all systems have in common* when they are simplified, or so well-blended together that the special atomic color effects get averaged out. This is called *thermal radiation*, and physicists define an idealized object that produces it as a *blackbody radiator*, or a *blackbody* for short. So a blackbody, by definition, when heated, emits only thermal radiation and has none of the alluring special color effects of sparkling fireworks. A blackbody is a physicist's idealization that everyday objects can only approximate, but they can approximate it pretty well. For example, though the sun is seen to emit light with strong absorption (Fraunhofer) lines due to the atoms in the surrounding cooler gas corona, the overall radiation is pretty much that of a hot blackbody. So, too, is a charcoal fire, as are the heating elements of the toaster, Earth's atmosphere, the mushroom cloud of a nuclear explosion, and the early universe. They are all reasonable approximations of a blackbody emitter of thermal radiation.

An excellent approximation of a blackbody is an old-fashioned boiler furnace, perhaps that of a steam locomotive, containing a hot coal fire. The furnace itself fills with approximately pure thermal radiation as it heats up. Indeed, this is what the late nineteenth-century physicists first studied when they wanted a good approximation of a blackbody radiator. To build a pure blackbody source, we need to isolate the thermal radiation from the coal fire. A hearty metal box that is large, durable, with thick walls, typically made of iron, and with a hole into which we could peer and insert instruments would suffice. We put the metal box in the furnace, allow it to heat up, and we peer into the hole. We are then looking directly at the pure thermal radiation that fills the cavity of the box. This radiation is emitted from the hot walls of the box and bounces around inside, and some of it comes out of the hole through which we are looking.

By peering through the hole we can study the thermal radiation to see how much of any given color (or wavelength) of light it contains. We can study how the color content changes as we increase or decrease the temperature of the surrounding furnace. What this amounts to is studying the pure radiation itself in thermal equilibrium.

If we raise the temperature of the blackbody furnace, we'll first only feel the gentle, warm but invisible infrared radiation emitted from the hole. At higher temperatures we'll be able to see a dull red glow of the light that escapes through the hole, like the wires of our toaster. Then, as it gets hotter still, the radiation will become bright red, then eventually yellow. If we have a powerful Bessemer steel manufacturing furnace (where oxygen is pumped in), we can go to extremely high temperatures and we'll observe the blackbody radiation as it becomes almost white in color. Then, if we imagine cranking up the throttle on our furnace to still higher temperatures (which we cannot achieve with any conventional furnace since it would melt), a brilliant bluish-white light—a mixture of all colors slightly tinted toward the blue—will stream out of the hole at the highest temperatures. We've arrived at the temperatures of a nuclear explosion or the surface of a hot bright blue star, like the blue supergiant,

Rigel, in the constellation of Orion—the most energetic emitter of thermal radiation within our part of the galaxy.[3]

Physicists have devised a way to precisely measure the intensity of the radiation emitted by the blackbody for various fixed temperatures. They also have discovered how to measure the amount of radiation at any given color, and they found that light of all wavelengths is contained in the blackbody radiation for any temperature, but some wavelengths are more prevalent than others. The fruit of these precise measurements was something called "the blackbody radiation curve," and its measurement was a difficult but heroic scientific achievement. We show a plot of the famous "blackbody curve" or "blackbody distribution" in figure 13.

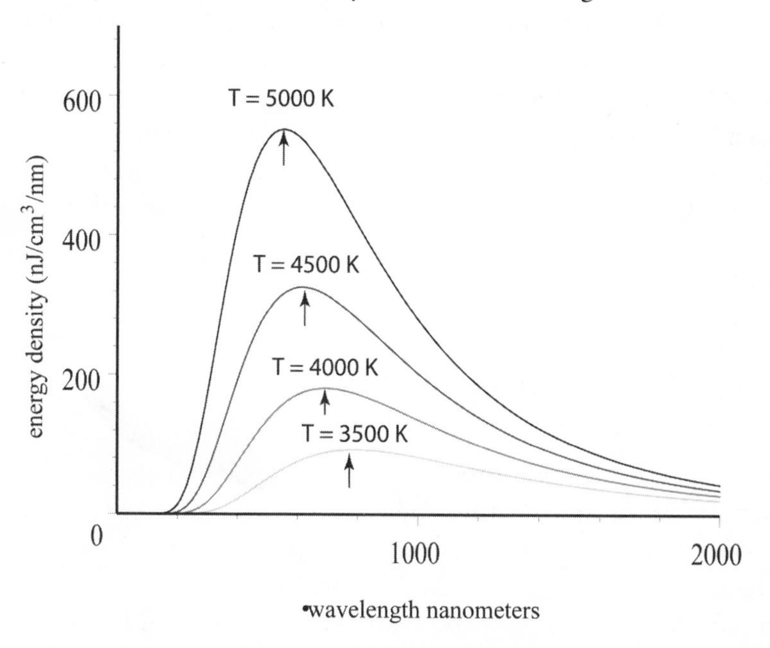

FIGURE 13: The blackbody spectrum of light radiated from a hot object at the indicated temperatures. The arrows indicate the peak wavelengths for each curve. Hence, for "lower" temperatures, T = 3500 K, the peak is near 800 nm, infrared light, and one would observe a red glow; for higher temperatures, T = 5000 K, the peak is moving toward the yellow near 600 nm. At much higher temperatures, the peak moves into the blue. Note that in all cases the energy in the shortest wavelengths is suppressed, as Planck explained with his formula $E = hf$. (The plot is the differential energy density of emitted light in nanojoules per cubic centimeter, per unit of light wavelength in nanometers plotted against wavelength in nanometers.)

The blackbody curve confirms our intuition about the progression of color with increasing temperatures. At lower temperatures in the furnace, 3500 K ("K" stands for "Kelvin"),[4] most of the emitted light has very long wavelengths, those of infrared and the deep reddish colors. As we raise the temperature, the peak of the intensity of light moves increasingly toward shorter wavelengths, or bluish colors. Many other wavelengths are all being emitted together so that the colors blend and we end up with a bright white light. Go to still higher temperatures, and the color would become whitish-blue (or bluish-white, if you prefer). It would get bluer still at higher temperatures, but now the wavelengths are in the ultraviolet range, which we cannot see.

The study of thermal radiation was a rich new subject that conjoined two fields of physics: the study of heat and thermal equilibrium—that five-syllable word, *thermodynamics*—and the study of light in radiation. The seemingly innocuous data from these areas yielded some interesting physics for study, but no one realized that these were clearly important clues into what was to become the detective thriller of the millennium—the quantum properties of light and the atom (after all, it's the atom that's doing all the work).

ICH BIN EIN BERLINER

Physicists in the nineteenth century, particularly a brilliant group in Berlin, spent much time heating up blackbodies and drawing the precise curves of the light intensity they gave off at each value of wavelength. With remarkable ingenuity they devised instruments to take a "spoonful" of emitted wavelengths, say, between 652 nanometers and 654 nanometers (see chapter 3, note 4 and figure 12 on wavelengths of light) in the red region and to quantify the intensity of the radiation in that band. Once they knew the numbers, they could look at a graph of the radiation from a blackbody and instantly name the temperature.

When it comes to a blackbody and temperature, you needn't worry about the details of the particular hot radiating object because all blackbodies, or things that approximate them, emit the same shape of the radiation curve at any temperature. But what you need to know about the reams of data that were accumulating by 1900—the shape of the blackbody curves for any temperature—is that (drumroll): *they all looked wrong*! They were *inexplicable*. The results simply did not compute, if you trusted Maxwell's equations and the well-honed laws of thermodynamics (which, if you were a turn-of-the-century physicist, you did), and the brilliant computational skills of a theoretical physicist named Max Planck.

A powerful theory of heat and temperature had already been developed in the nineteenth century, known as *statistical mechanics*. This was devised by Maxwell himself and (at the time obscure) American theorist J. Willard Gibbs,[5] based on a mathematical formulation of the great Austrian physicist Ludwig Boltzmann (whose life had tragic proportions).[6] Maxwell's, Boltzmann's, and Gibbs's theories taught us how to calculate the ways that the many different components of a system should move, that is, how their motion is distributed, when the system is in thermal equilibrium. Max Planck, sewing these ideas together with Maxwell's triumphant theory of electromagnetic waves (and it is complicated), reasoned that he could compute the very shape of the blackbody curve.

Planck discovered that the curve should have exactly the shape that is seen in experiment for the longest wavelengths of light emitted from a blackbody. But he found that the curve should blow up, that is, become infinite, for the shortest wavelengths of light (the ultraviolet colors). In other words, the blackbody curve should always be skewed toward the violet (shortest-wavelength) part of the spectrum for any temperature. This latter effect is in gross disagreement with the experiments.

To put it another way, according to Planck's careful calculation, a "spoonful" of short-wavelength (blue-violet) radiation should always have far more intensity (brightness) than an equal "spoonful" of lower-frequency (red) radiation. This happens essentially because blue light is "smaller" (it has a shorter wavelength), and you can squeeze more of it into a given amount of space. Planck thus predicted that, in Maxwell's classical theory of light, all hot objects should therefore always brightly shine bluish-white at any temperature. But at low temperatures, by experimental observation of the blackbody curve, there was far more red than blue. In fact, at low temperatures there was essentially no blue light at all.

So what's going on? An entertaining metaphor may help here. Suppose we have an auditorium in which, for a special fixed low admission price, you can sit anywhere and hear a famous pianist, Alfred Brendel, perform Beethoven's Piano Sonata no. 15. The audience is composed of amateur musicians who are all *extremely* thin and very friendly. So how does this audience of ultra-anorexic piano aficionados arrange themselves throughout this auditorium? Remember, they can sit anywhere, at the same price. Any guesses? You probably got it right. As these listeners love great performers, Brendel, Beethoven, and particularly Sonata no. 15, all two thousand listeners crowd down to the left corner, nearest the pianist and the piano keyboard, all sitting in the same seats (remember, they're thin enough to fit), with just a handful of the more musically astute (who want to hear the piece rather than watch it) spread out over the rest of the auditorium. The music amateurs want to behold the pianist's hands as they fly and dance over the keys of the Steinway concert grand piano. So how is this a physics metaphor? This scenario corresponds to the prediction coming from the classical theory of thermodynamics and light: there is a strong preference for produced blackbody radiation to crowd toward the smallest wavelength, the bluish-most colors. Smaller wavelengths, simply being smaller, allow more crowding together than their long-wavelength counterparts.

But, of course, this is not what happens, neither in the music world nor in blackbody radiation. At an actual concert, the front-row seats are very expensive and are often sparsely occupied, and the back rows and highest balconies are empty (where you often can't see or hear a thing!), with the bulk of the audience ensconced in the middle rows. Analogously, the observed distribution of intensity of blackbody radiation begins small (at long wavelengths), builds up to a peak at a wavelength that depends on the temperature, and tails off at the very short wavelengths. There is in nature no observed crowding of light into the short wavelengths at all. In fact, the ultra-short wavelengths are strongly suppressed. Yet, as Planck observed, Maxwell's theory, using Gibbs's and Boltzmann's ideas about thermal systems, predicted that the crowding into the blue should happen. But it doesn't. So why doesn't it?

CATASTROPHE! (IN ULTRAVIOLET)

The classical theory of light, according to Planck's calculations, had predicted a wavelength (or color) distribution that zoomed up and up as the wavelengths got smaller. In fact—much to the exasperation and perplexity of the theorists—the theory predicted infinite intensity at tiny wavelengths, such as in the far ultraviolet. Someone—perhaps a newspaper reporter—called this situation the *ultraviolet catastrophe*. It was a catastrophe because the supposed preference for the ultraviolet doesn't occur in the data. In fact, if it did, fires wouldn't glow red at lower temperatures, a fact known for hundreds of thousands of years—they would glow blue.

So here was one of the early cracks in the heretofore triumphant classical physics. (Gibbs had found another, and perhaps the first one, some thirty-five years earlier, whose significance was not appreciated at the time, except possibly by Maxwell.)[7] The point is, the data begged to differ from the classical theory. The blackbody curve (figure 13), depending on the temperature, shows a peak at some wavelength that depends on temperature (in the red at low temperatures and the violet at high temperatures). Then the curve drops rapidly at the still shorter ultraviolet wavelengths. Now, what happens when a beautiful and well-honed theory, created by the greatest minds of the century and confirmed by the authorities in the science academies of Europe, collides with some grubby, ugly facts of reality? In religion, sheer dogma is perpetual; in science, a bogus theory goes out with the morning trash bin.

Classical theory predicts that your toaster wires should glow blue, but instead, the glow is a dull red. Thus, whenever you gaze into the depths of your toaster, you are seeing a phenomenon that is in violent disagreement with the expectations of classical physics. Moreover, while you may not appreciate it yet, you are also seeing direct evidence that light comes in lumps—in *quantum particles*. You are witnessing quantum physics firsthand! But, you protest, did we not show, by way of the genius of Mr. Young, in the previous chapter, that light is a wave? Yes, we did, and it still is. So we must now prepare for things to get a little weird. Remember, we are voyagers to a strange new world far, far away—yet we can find it in the glowing wires of our toaster.

MAX PLANCK

Back in Berlin, at the center of the ultraviolet catastrophe, presided Max Planck, a forty-year-old theoretical physicist at the University of Berlin who was an expert on the theory of heat.[8] Planck was fully aware of the ultraviolet catastrophe and wanted to understand what was going on. In 1900, poring over the blackbody data that had been generated by his Berlin colleagues, Planck was led to a mathematical trick that, using Maxwell's, Boltzmann's, and Gibbs's ideas of thermal physics, produced a formula for the blackbody curves that turned out to be in excellent agreement with the experimental data. Planck's trick allowed the long-wavelength red light to be copiously radiated at any temperature, more or less just as in the classical theory, but his mathematical gimmick effectively charged a "toll" for the emission of the shorter-wavelength light. This penalty for emission of short-wavelength light suppressed the bluish light (remember: short-wavelength = higher frequency = blue), so blue would be less copiously radiated.

The mathematical gimmick seemed to work. Through Planck's "toll," the higher frequencies "cost" a lot more energy to be produced, while the lower frequencies cost less energy to produce. Then, Planck correctly reasoned, there wouldn't be enough energy, at a certain temperature, to excite the short wavelengths. In our concert hall metaphor, Planck had found a way to depopulate the front rows and get more people to go into the galleries and higher balconies: he simply charged more for the main floor seats and much less for the balconies. In an uncharacteristic burst of inspired insight, he related the wavelengths, or equivalently, the frequencies, of light to energy: the shorter the wavelength, the higher the frequency, and, so said Planck, the higher the energy.

This may seem like a simple idea, and in many ways, thanks to nature, it is. However, the classical theory of light made no such prediction at all. The energy content of light in Maxwell's classical theory depended only on the *intensity* and not on the color or frequency. So how could Planck sneak in this extra aspect of the light in the blackbody spectrum? How could he impose the idea that energy depends on intensity as well as frequency? There is still a missing element here unless you specify, "what" exactly has higher energy when it has higher frequency.

To solve this problem, Planck effectively divided up the emitted light at any given wavelength (frequency) comprising the blackbody curve into bunches, or *quanta*, allocating to each "quantum" an energy that directly related to its frequency. Planck's inspired formula is actually the simplest one possible:

$$E = hf$$

or in other words, "the energy of a quantum of light is directly proportional to its frequency." What this means is that electromagnetic radiation comes out in lumps such that each "lump" has an energy equal to some constant, h, times the frequency. The total intensity of light at any given frequency is *the number of quanta* that are being detected at that frequency times the energy of those quanta (recall that frequency is inversely related to wavelength). In Planck's effort to fit the data, high frequencies, which are equivalent to short wavelengths, cost more energy to be emitted in a blackbody. When you plugged in his equations and dialed the given temperature, the predicted blackbody curve agreed precisely with the data.

Remarkably, Planck did not really see his modification of the successful Maxwell theory as pertaining directly to light. Rather, he envisioned that it really pertained to the atoms in the walls of the blackbody radiators, that is, as to how the light is emitted. The penalty of emitting blue light over red wasn't viewed as an intrinsic property of blue light versus red light but rather as a property of the way atoms dance around and radiate a given color of light. This way Planck hoped to avoid potential conflicts with the Maxwell theory that had otherwise worked so perfectly. After all, the direct connection of light to electricity and magnetism was established in Maxwell's theory. Moreover, electric motors were driving trolleys around European boulevards, Marconi had already invented radio telegraphy, and people were designing sophisticated antennas. Maxwell's theory wasn't obviously broken, so Planck didn't want to fix it. He preferred to "fix" the more arcane theory of thermal physics.

Yet here were two dramatic departures from the classical theory, at least the theory of thermal radiation. One was to connect the intensity (energy content) of the radiation to frequency (which was completely absent in the Maxwell theory); the other was to introduce lumps, or discreteness—the "quanta." These are logically intertwined. For Maxwell, the intensity is smooth and can have any continuous value, dependent only on the amplitudes of the electric and magnetic fields in the wave disturbance of light. In Planck's treatment, intensity at a certain frequency becomes the number of quanta in the light at that frequency, where each quantum carries an energy proportional to its frequency à la $E = hf$. The new idea of quantum lumps is starting to "feel" more like the concept of particles, yet light is seemingly still very much a wave, according to all the experiments on diffraction and interference.

But no one, including Planck himself, really understood the full significance of this breakthrough. Planck thought of the quanta, each with its ration of energy equal to hf, as short bursts or pulses of radiation, coming from the thermally agitated atomic motion in the blackbody walls, somehow induced by the unknown details of the thermal emission processes of the blackbody. He did not foresee that his constant, h—now known as *Planck's constant*—would become the cornerstone of the coming revolution, heralding the birth of the baby quantum theory and the modern era. Incidentally, Planck's great discovery was made at the age of forty-two. In 1918, Max Planck was awarded the Nobel Prize for his discovery of the "quantum of energy."

ENTER EINSTEIN

The awesome implications of Planck's quanta were first appreciated by none other than the young, still obscure Albert Einstein, who, after reading Planck's paper in 1900, exclaimed, "It was as if the ground has been pulled out from under me."[8] The central issue was whether the quantum lumps were a product of the emission process that produced the light or an essential property of light itself. Einstein understood that Planck had introduced something disturbingly sharp, discrete, or particle-like, into the emission process that created the light from heated substances, though Einstein initially would shy away from the revelation of seeing this lumpiness as a fundamental quality of light itself.

A few words about Einstein are due. An un-precocious child who disliked school, Albert Einstein would not have been voted "most likely to succeed" by anyone. But he had been bewitched by science since the age of four, when his father showed him a compass and he became mesmerized by the invisible forces that caused the iron needle to be drawn unfailingly toward north no matter which way the compass was rotated. In his seventies he would write, "I can still remember—or at least I believe I can remember—that this experience made a deep and lasting impression on me." A few years later, the young Einstein fell under the spell of algebra, which his uncle had taught him, and was deeply moved by a geometry text he read at age twelve. He went on to author his first scientific paper, on ether in the magnetic field, when he was sixteen.

When we meet him here, Einstein was still living in obscurity. Unable to obtain a regular academic position after graduation, he worked intermittently as a tutor and a substitute teacher before landing a job as an examiner at the Swiss patent office in Berne, Switzerland. Though he had only weekends free for his own research, his seven years in the patent office were the ones in which he laid the foundations of twentieth-century physics, including showing how to count atoms (how to measure Avogadro's constant), inventing the special theory of relativity with its profound consequences for our understanding of space and time and $E = mc^2$, and contributing to quantum theory, in addition to other achievements. Among his other talents, Einstein had the gift of synesthesia, a phenomenon in which one sense—vision, for instance—conjures up another, such as sound. When he was working on a problem, his mental processes were accompanied by visual images, and he always knew when he was on the right track because he felt a tingling in his fingertips. Einstein would not become a household name until 1919, when phenomena associated with a solar eclipse confirmed his general theory of relativity. However, it was for his explanation of the photoelectric effect in 1905 (and not special or general relativity!) that he won the Nobel Prize.

Try to imagine the culture shock in 1900. There you are, serenely reviewing data—studies of the continuous spectra radiated by heated objects—that were written since the middle of the nineteenth century. These were experiments on electromagnetic radiation, which was well understood as waves since the 1860s, thanks to Maxwell. That this quintessentially wavy stuff could behave in special circumstances as if it were bundled energy packets, in other words, "particles," threw the classical community into a monumental state of confusion. But Planck and most of his colleagues assumed that some sensible, essentially neoclassical, über-explanation would be forthcoming. After all, blackbody radiation was a complex phenomenon, like the weather—many simple-to-understand things often conspire to make complex things seemingly incomprehensible. Perhaps what is most incomprehensible about all this is that nature was now truly revealing to patient observers her innermost secrets.

THE PHOTOELECTRIC EFFECT

That low rumble you hear in the background is the foundation of classical physics—the brainchild of Galileo, Newton, and Maxwell—beginning to crumble. The next tsunami to hit it was something called the *photoelectric effect.* When you snap a photo with your cell phone, you are using the photoelectric effect (incarnated in photocells). The basic principle is this: "light goes in, electricity comes out."

The German physicist Heinrich Hertz first observed the photoelectric effect in 1887 when he noticed that if a polished metal surface is struck by light it will emit electric charges; in other words, electrons will pop out. But this happens not with just any light. It must be short-wavelength (high-frequency) light. Red (long wavelength, low frequency) won't do the trick; only violet will. Sound familiar? Is there a connection here?

Before we discuss what Einstein did about the photoelectric effect, let's back up a bit and focus on the electron, discovered in 1897 by J. J. Thomson of the respected Cavendish Laboratory at Cambridge University. The little electron is a "corpuscle" of electric charge, a pointlike particle with no internal structure but with a mass, only about 1/2000th of the mass of the atom. (The past century of beating up on the electron with "atom smashers" of ever-increasing power has not succeeded in breaking up the electron into anything smaller, so as of this writing, we have concluded that the electron is a truly fundamental particle, irreducible and structureless.)

At the turn of the nineteenth into the twentieth century, electrons were known to play a crucial but not yet clearly understood role in the makeup of atoms. And, for our purposes, it is important to know that when light shines on a metal surface—especially if it is a highly polished, good electrical conductor (grease, grime, and oxidation on the surface of the metal will interfere with the result)—out come electrons. That's the photoelectric effect. In goes light—out comes electrons!

So here's what happens. Think of a beam of light whose intensity (brightness) and color (wavelength or frequency) we can change. We make this light shine on a clean metal surface. We find that when we shine dim red light on our metallic surface, nothing happens. We make the red light more intense, and apart from things getting a little warm, it produces nothing. Now we make the light dim but we change the color to blue or violet (shorter wavelength, higher frequency). Suddenly we observe when we shine this blue light on the metal that a few electrons come popping out of the polished surface. We next increase the intensity (brightness) of the light, and now many, many electrons rapidly pop out of the surface.

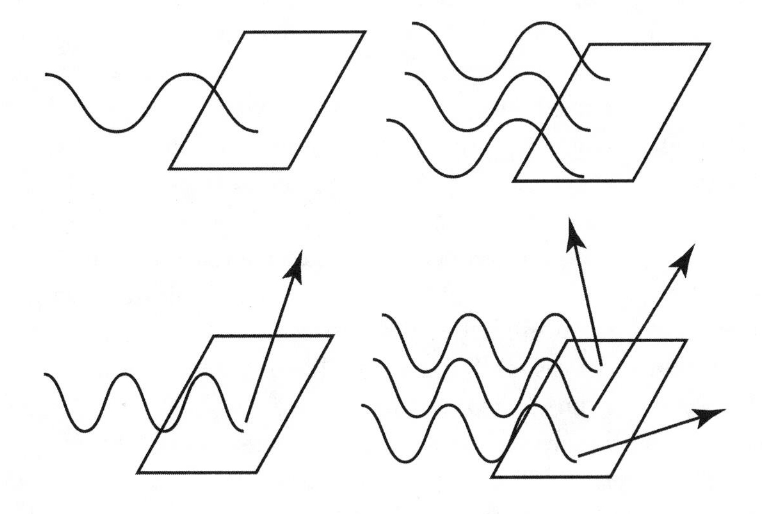

FIGURE 14: The photoelectric effect. As explained by Einstein, dim (red) light has insufficient energy per photon to eject electrons. Likewise, bright red light has insufficient energy per photon to eject electrons and will simply heat the material up. Once the light has a sufficiently short wavelength (blue), corresponding to energetic photons, then a few electrons are ejected. Bright blue ejects many electrons. This directly reveals the energy content of a beam of light is $E = Nhf$, where N is the number of photons and f the frequency per photon (i.e., hf is the energy per photon).

We have observed something remarkable. The emission of electrons doesn't depend only on how bright the light is; rather, it depends critically on the color of the light. There is a *threshold frequency* of the light, below which no electrons emerge (i.e., we must have light that has a short enough wavelength, shorter than a certain value, corresponding to the threshold frequency). For light below this threshold frequency (red light), we get absolutely no electrons—nada—zilch. It doesn't help if we make the red light very, very bright or just leave it on for a long time. No electrons emerge. This is actually quite odd because, according to Maxwell's classical wave theory, the bright or intense light should mean a lot more energy. Also, according to Maxwell's theory, the energy content of light does not depend on its frequency, just on its intensity or brightness. The photoelectric effect, the ability to knock electrons out of a metal, depends critically on the frequency (or wavelength) of light. And when we do get above the threshold frequency (or below the threshold wavelength), we do get more electrons as we increase the brightness of the light, in accordance with Maxwell.

Normally electrons are trapped in the metal surface, tied somehow to the array of atoms. To "kick" an electron out, we would need to give it enough energy to get through the surface, so it could go its merry way (and clean surfaces make it easier). So as we double or triple the wattage of our light source, we get no electrons as long as we stay below the frequency threshold. Astonishingly, though, when we raise the light frequency (decrease the wavelength) to get just above the threshold, out come electrons—even if the light is weak, even if we have replaced our thousand-watt bulb with a little ten-watt nightlight. And the electrons come out instantly.

Fortunately, it is pretty easy to measure the energy of the electrons that pop out of the metal: a well-equipped high school physics lab can repeat these historic experiments that took place in laboratories from Milan to Berlin and Stockholm from the late 1800s onward. Experiments in science are always repeatable and must produce the same results at any time and any place (unlike appearances of ghosts at séances or religious icons on English muffins). And, sure enough, it turns out that the energy of the emitted electrons always depends only on the frequency (or wavelength) of the light, and not on its intensity. Once we have electrons emitted from the metal, raising the intensity changes the number of electrons emitted per second but does not affect the energy of each electron.

But raising the frequency of the light, that is, decreasing the wavelength, say, from blue-violet's 4500 Å to violet's 3500 Å, does boost each of the electrons' energy. The labs all over Europe recorded the same mystical frequency thresholds, and these data were obviously giving correct descriptions of nature—at least the part of nature involving light and metal surfaces and electrons.

What should we make of it all? Basic to any explanation of the photoelectric effect is the concept that there is an energy toll to be collected at the surface of the metal that must be paid by the escaping electron. If the electron doesn't have enough "money" (energy), then it can't pay the toll and it won't get through the barrier. The classical wave theory of light, which had prevailed since the early 1800s, simply couldn't cope with the data. The energy in the electromagnetic wave depends on the amplitude of the wave (its magnitude from trough to crest). Yet the data insist that even huge intensities of waves are useless in releasing electrons if the frequency is too low. And, even above the threshold frequency, the classical wave theory would suggest that a low-intensity wave—which is spread out over a huge number of atoms—would find it extremely difficult, if not impossible, to concentrate all its energy on a single electron in a very short time to make it "pop out." Finally, why does the emerging electron energy depend on the frequency of the light that emitted it at all? The classical theory of light makes no such connection.

Okay, it's now up to a Sherlock Holmes to assemble the evidence and come up with some sensible explanation (stop and contemplate what you have read to this point and perhaps you can guess the answer). In fact, it was Albert Einstein, in 1905, recovering from the trauma of his doctoral exams and stealing some time from his job as a patent examiner in a small but clean office in Berne, Switzerland, who hit upon the solution. He remembered Planck's essay on the blackbody radiation and wondered: If light can appear to be composed of lumps, or quanta, in emission, couldn't the same thing be happening in the absorption of light? Could the energy of the light actually be concentrated in lumps, and then be proportional to the frequency? Recall Planck's formula for the emission of quanta in blackbody light: $E = hf$, the energy (E) being equal to the frequency (f) times a number, h, which is always the same. Suppose, mused Einstein, what Planck's formula really describes is not just a complex emission problem in thermodynamics but what light really is. If this is the only way light comes—in lumps or quanta—then its energy can be offered to the electron, as an all-or-nothing proposition, in a direct collision between the electron and a light quantum itself. The electron swallows this quantum of energy. If the energy of the swallowed light quantum is bigger than some threshold amount—call it *W*—then the electron has enough energy to pay the exit toll from the metal and escape through the surface. This means the light quantum must have a threshold frequency—call it *F*—so that hF is bigger than or equal to *W* to cause the electron to escape from the metal. The electron has enough energy to escape the metal as long as the light's frequency is greater than *F*—just like the data say. Blue can do it. Red cannot. This elegantly simple idea completely explains all the logic of the experimental data on the photoelectric effect.

This explanation of the photoelectric effect won Albert Einstein in 1922 his only Nobel Prize. Einstein gave us a new interpretation of Planck's idea in the quantum property of light. Light quanta are not just some complex mechanism associated with emission or absorption in the thermal blackbody walls as Planck had thought. Rather, the quantum property is intrinsic to light itself. Soon afterward the quanta of light began to be called *photons*. Indeed, light is composed of photons, which are particles and are dealt with like any other particles in the laboratory. The energy of each photon is proportional to its frequency à la Planck's formula, $E = hf$. High-intensity light, in Einstein's view, that ejects a large number of electrons from a metal's surface must correspond to a huge numbers of photons. But, to eject electrons from a metal, each photon must have an energy, hf, greater than threshold *W*. If the photon energies are less than *W*, then none of the electrons can escape through the metal's surface.[9]

Over the next several years, this theory would be tested carefully by dozens of experimental scientists. It must have been correct since it gave us television. One can now look up the energy toll paid by an electron in escaping for any metal. It is what we call the "work function of the metal," or *W*, which is tabulated in reference books. *W* depends on the atomic structure of the metallic substance. Substances found with low values of *W* make it easier for electrons to escape the metal. They are therefore used to coat the surfaces of photocells, making them more efficient. Solar cells, now in great abundance, supply electrical power for people's homes and factories. Since sunlight can be readily converted to electric current, solar cells producing raw electricity for our consumption will become a significant part of the solution to our energy crisis. Today a large part of "nanotechnology" is based on the creation of nanoscale (sizes approaching large molecules) photoelectric devices called "quantum dots" that can emit an electron of any desired energy by eating a photon, and vice-versa. In addition to more efficient solar cells, quantum dots may have profound medical applications as in attacking cancer cells with energetic electrons.[10]

So here is the situation after Einstein's explanation of the photoelectric effect: The concept of electromagnetic energy propagating as waves, extending continuously throughout space, explains various phenomena, such as reflection, refraction, diffraction, and interference. But in the case of blackbody radiation and the photoelectric effect, the wave theory failed, whereas a particle/quantum model succeeded. In this latter picture of the properties of particles, each particle (photon) carries a definite quantum of energy, with Planck's formula $E = hf$.[11]

ARTHUR COMPTON

In 1923, the photon-as-particle began to take on even more significance when Arthur Compton, with a Princeton PhD, began to study how x-rays (very short wavelength light) behave when they strike electrons. His results were clear: photons in collision with electrons behave just like particles and collide with electrons just like billiard balls.[12] The electron, initially stationary, recoils in the collision like a billiard ball. But the photons themselves are also seen to recoil in the collision. The collision of an energetic photon with an electron, yielding a recoiling photon and electron, is a process now called "Compton scattering."

This process is like any other collision process in physics, conserving the total energy and momentum of the electron and photon. But it cannot be understood without assigning a definite particle property to the photons. Compton did not arrive at his radical explanation for the scattering process until all his previous attempts at an explanation had failed. By 1923, the first effort at constructing a quantum theory, the "old quantum theory" of Niels Bohr, could not provide a basic explanation for the Compton scattering process. It would require the development of the new quantum mechanics to explain its significance. And when Compton reported his discovery at meetings of the American Physical Society, many of his peers were downright hostile.

Compton, the son of hardworking Mennonites from Wooster, Ohio, persevered, fine-tuning his experiments and interpretations. The last public debate on the topic was held in Toronto in 1924 at a meeting of the British Association for the Advancement of Science, at which Compton spoke persuasively at a specially convened session. Then his sometime nemesis, Harvard's William Duane, who had previously failed to confirm his data, returned to his lab and personally repeated the disputed x-ray experiments. He now conceded that the "Compton effect" was valid. Arthur Compton was awarded the 1927 Nobel Prize, played a major role in the development of twentieth-century physics in the United States, and made the cover of *Time* magazine on January 13, 1936.[13]

So what does all of this show? We have a series of phenomena indicating that light is a stream of particles—light quanta called photons. (Ah, Newton, how could you have known?) But we also have Young's double-slit experiment proving the wave theory of light (and the millions of repeat experiments corroborating that in high school and college labs the world over). The three hundred-year-old conundrum thus returned. Had physics reached an ultimate paradox? How can something be both a particle and a wave? Must we give up on physics altogether and take up Zen motorcycle maintenance?

THE DOUBLE-SLIT EXPERIMENT RETURNS WITH A VENGEANCE

How can various phenomena indicate two contrary views of light? Are there perhaps two kinds of light? Perhaps the wave and particle properties are not contrary? Let's see: light waves can be everywhere in space; particles are always at specific locations. Waves can be subdivided—for example, 80 percent reflected, 20 percent absorbed by a surface. A particle cannot be subdivided; it is or it isn't there. But the most crucial difference is still illustrated by the quintessential double-slit experiment, first invented by Dr. Young.

To explore this paradox further, let us follow a series of experiments, all of which have been carefully performed many, many times. We'll begin by repeating Young's experiment using monochromatic light—say, pure blue light of a definite wavelength—shining on a screen with two small horizontal slits. Beyond this screen is another screen, a "detector screen," covered (at first) with photographic film. We turn on the light source for, say, a few minutes, then develop the film. There we see an "image" of the slits displaying vertical stripes, "interference fringes," exactly where they are supposed to be (as in figure 8).

Since we are impatient with the slow process of developing film, we'll use modern technology to replace it with thousands of tiny light detectors that detect the presence of light instantaneously. These are photocells that each (via the photoelectric effect) give rise to a current of electrons when blue light shines on their surface, and each photocell current is read out on a meter. The photocells are arranged like a mosaic of small tiles on a bathroom wall, covering the detector screen.

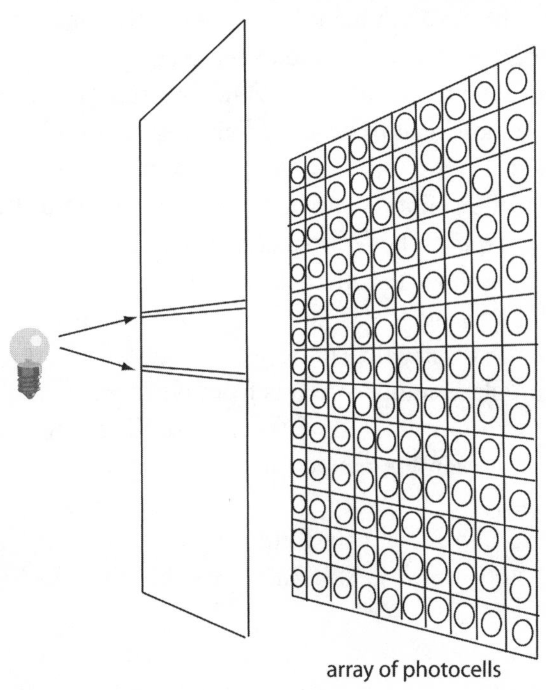

FIGURE 15: A Young's interference experiment designed to count individual photons with an array of photocells.

Now we turn on the light that shines on the screen with two slits. Some rows of meters, corresponding to some rows of photocells, read high; other rows read zero; other rows in between read an intermediate number. The high current rows match exactly the bright stripes we had just seen on the photographic film. So photocells also work just fine in showing the interference pattern, and we don't even have to develop them with nasty chemicals or wait. As you may have guessed, this light observed with photocells shows wavelike interference: the waves arriving from each slit reinforce one another in some places and cancel out in others. We can also show that if we close one slit, the pattern dutifully disappears (as in figure 9), replaced by a broad smear of currents, the peak of which appears opposite the remaining open slit. So, as Young demonstrated, we need both slits open to generate the cancellations and reinforcements of the interference patterns.

Now we do something new. We make the light source very dim. If it's an electric lightbulb, we simply reduce the voltage. We repeat the experiment. The photocell currents now exhibit jerky responses, the meters swinging up, then down to zero, then up a little, then zero again. From experience with photocells, we understand this behavior. When we turn down the intensity, we're detecting the individual particles of light—we're observing *individual photons*. We can actually count the individual photons as they arrive at each photocell. To make the counting easier, we can automate this by arranging for each photocell output to be fed to a computer, such that each impact of each photon adds a number to a particular entry in our counting program for that photocell. So, using a very dim light source, let's reset all the entries to zero and begin. We display the counts for each photocell on a monitor as they begin to accumulate numbers of hits from photons.

We wait patiently until some of the photocells show numbers like 100 or more, and we examine the results of our photon counting experiment. Sure enough, the highest registers correspond to the photocells in the rows where we had previously found high currents that also corresponded to the bright lines on the photographic film. And we observe that there are rows of zeros—the dark bands of the interference pattern where no photons have arrived—the cancellation by interference. So even for very dim light with individual photons being "counted" by the photocells, we see the wavelike interference pattern. Now, that is surprising. Are individual photons interfering with each other, we muse?

So we reason that if we make the light source, our photon source, extremely dim, so that only individual photons are passing through the slits, one at a time, we won't see any interference pattern since now the photons cannot interact with each other. We take a deep breath. We turn the source of light down so low that we get only about one photon detected by a photocell per second (dimmer still and we could make it one photon per minute, or per week, or even per year if we had the time). After an hour of listening to the "tick, tick tick" as individual photons pass through our slits to the detecting photocells and being counted by our computer, we finally take a look at the data. We have, so far, collected only a few photons and we see an apparently random assortment of numbers spread out over the thousand or so photocells (see figure 16). It's low statistics . . . let's keep the experiment running longer.

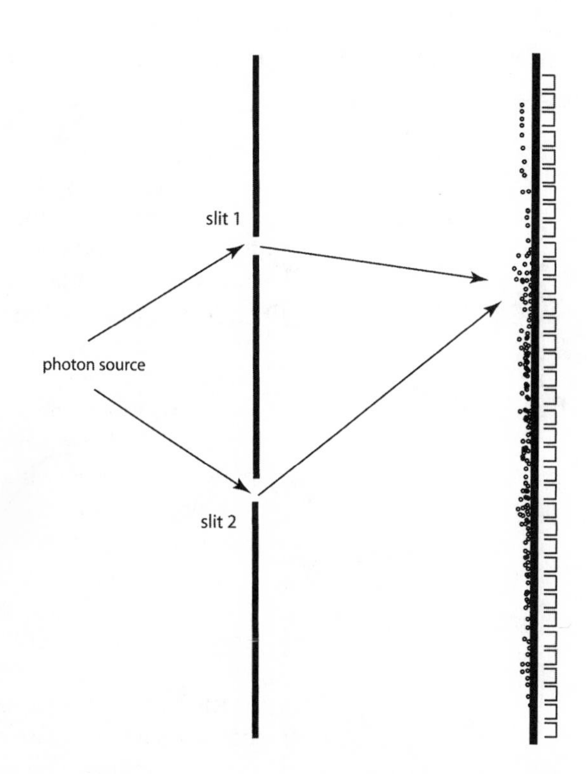

FIGURE 16: We count a few photons with the automated experiment and plot the output. So far we have limited statistics to see any pattern.

But as we run the experiment another few hours, a pattern in the data gradually emerges. We examine a row of photocell data—call it Row #6—with numbers like 67, 75, 71, 62, 68, and so on. Nearby Row #8 shows numbers like 33, 31, 26, 31, 28, 28, 27, and so on, while further along, our Row #12 reads 0, 0, 1, 0, 0, 0, 2, 0. Row #6 corresponds to one of Young's bright fringes in our interference experiment; Row #12 to one of the dark bands, where there are interference cancellations; and Row #8 to an intermediate place in the interference pattern. The interference pattern is back. But this time we are seeing the interference, not with continuous waves, but with individual particles—with individual photons, arriving once per second in our detectors. The photons are coming so slowly that they cannot possibly be interfering with each other. They are particles, but somehow they are interfering with themselves!

But maybe there's something wrong with our apparatus. Maybe the apparatus is faking an interference-like signal even if there isn't one there. Let's try to "turn off" the interference pattern in our experiment to make sure it goes away.

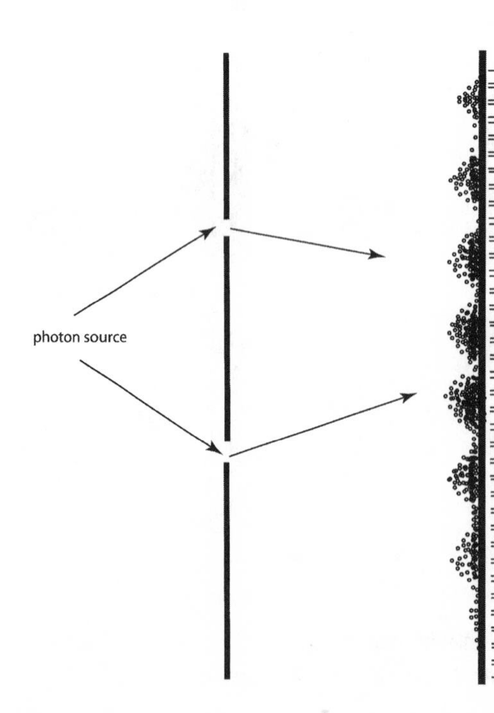

FIGURE 17: After counting many photons, one at a time, we see the familiar pattern. The photons pile up in the areas of the "bright bands" in Young's interference pattern, while few photons are detected in the "dark bands." This confirms the fact that the intensity of light is determined by the number of photons, but now we have a new mystery: how do individual photons interfere with one another? Even if we do the experiment with a very dim light source and count the photons very slowly, perhaps only one photon per hour, we still get the interference pattern. Therefore, this is not interference between two or more photons. Single photons, one per year, will produce the pattern if we wait long enough to collect a large sample of them.

With trembling hands, we close off one of the two slits, reset our counters, and turn on the very dim light source. Slowly a pattern of counts forms: a broad swath with the highest numbers opposite the open slit. Notice especially that our Row #12 now reads numbers like 21, 20, 17, 18, 20, 19, 15 . . . , whereas, with both slits open, it read 0, 0, 1, 0, 0, 2, 0 . . . . By closing one slit we have successfully turned off the interference effect. So what does this mean? This completely reproduces what Young observed for continuous waves, though we are simply counting discrete particles. The implications should cause goose bumps on your arms and the hair to stand up on your neck, and various other titillations.

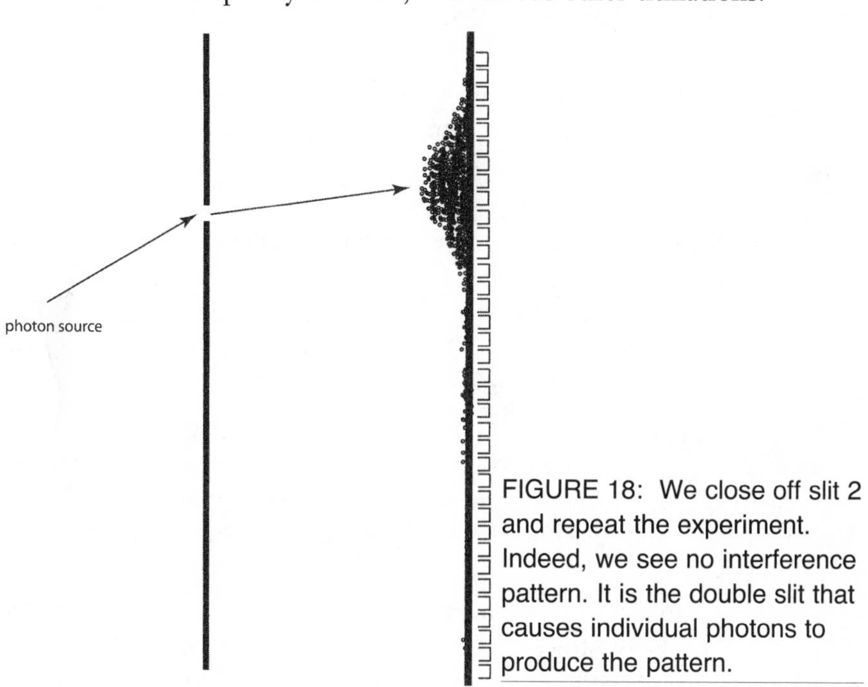

FIGURE 18: We close off slit 2 and repeat the experiment. Indeed, we see no interference pattern. It is the double slit that causes individual photons to produce the pattern.

A single photon seems to "know" if there are two slits open or only one slit open and it behaves differently accordingly!

With what is a single photon, a single corpuscle of light, interfering? Along what path through the two open slits did it take? How does a single photon know there is another path available that it didn't take? Yes, dear reader, this strange phenomenon really happens. The experiments have been repeated in many guises over and over many times. The outcome of an experiment involving the motion of a pointlike particle depends on all the paths the particle *might have taken* as well as the particular path it actually does take. This may be the most spooky thing we can directly observe about reality. Nature, the physical universe we inhabit, is evidently a haunted house.

BOOBY-TRAPPING THE SLITS

We have now witnessed a dramatic confrontation between wave and particle behaviors. A single photon, the energy quantum of light, passes through slit 1, but if slit 2 is open as well, it experiences "interference," just like Young's experiment predicts for waves. If slit 2 is closed, on the other hand, there is no "interference," also like Young's experiment predicts.

But a photon *is a particle*. Surely the particle goes through either slit 1 or slit 2. If it goes through slit 1, how does the particle/photon "know" whether slit 2 is open or closed so it can experience interference? The only possible explanation is an absurd one: that the photons passing through slit 1 somehow "know" whether or not slit 2 is open and, when it is, change their trajectory to avoid hitting certain of the photocell detectors, the ones that correspond to interference cancellations. In other words, these spooky light particles are feeling out both slits to determine how many are open before they "decide" where to land. Is that not absurd?

Obviously, this is a pretty bizarre thesis, but we can test it: We can try to watch the tricky photons to see which slit they pass through by placing a photon detector behind one of the slits, let's say slit 1. This single detector is like a highway patrolman seated on his motorcycle behind a billboard. It will simply alert us whenever a photon passes through that slit. This lone detector is sufficient to determine whether any given photon went through slit 1 or slit 2 (since any photon we end up counting that is not seen by the detector must have gone through slit 2).

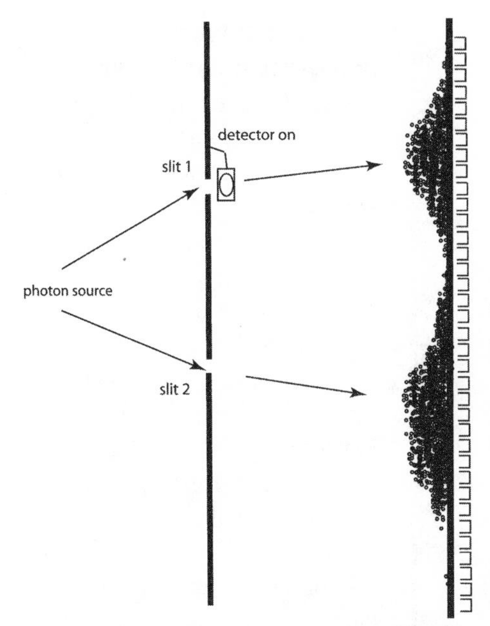

FIGURE 19: Now we try to fool the photons with a "booby-trap." We place a detector at slit 1 to record if a given photon passed through this slit or not. With the booby-trap detector "on," we repeat the experiment. We do not see the interference pattern, but just two pile-ups of photons under the two slits.

Now we redo the experiment. We use dim light or bright light, but we have our new slit detector alerting us every time a photon goes through slit 1. We are simply "tagging" which slit any given photon went through. And after a while we check the results: the interference pattern has vanished (figure 19). So now we turn off the slit detector and make no measurement as to which slit the photon passed through. The result: the interference pattern is back (figure 20). We repeat the experiment many times, with detector on and detector off. We even place the detector on slit 2 instead of slit 1. What we discover is that whenever we observe the slits, to see which slit each photon went through, we destroy the interference pattern. When we don't observe which slit the photon went through, voila! we again get the wavelike interference pattern. Is this an accident? A clumsy experiment? Is something really eerie going on? Perhaps making an observation at the slit counter does something to the photon's path, messing up the interference pattern. This is not unreasonable; it wouldn't take much to knock the photon off course. It does seem as if nature is deliberately frustrating our efforts to understand how particles can create an interference pattern.

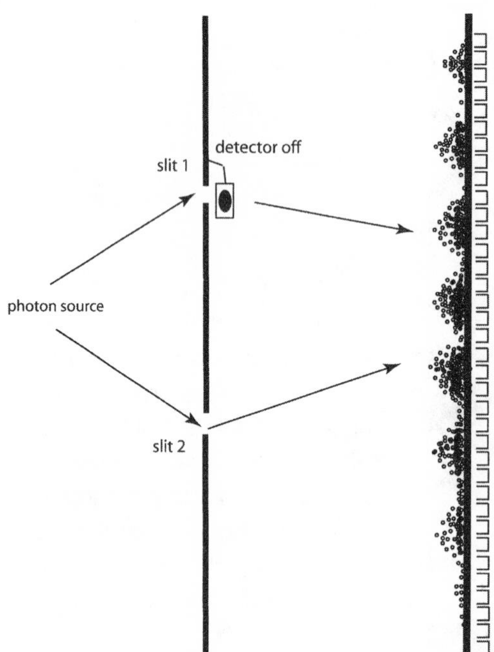

FIGURE 20: We turn off the booby-trap detector and repeat the experiment. Now no one, or no thing, is watching which slit the photon went through. Again we get the interference pattern. The mere act of watching which slit the photon passes through destroys the interference patter and yields the result of figure 19. If no one (or no thing) is watching, we get the interference pattern. This works with photons or electrons (or all other elementary particles, even individual atoms).

As a final check we place one slit detector at slit 1 and a second one at slit 2. Again, whenever we register which slit the photon went through, we destroy the interference pattern. When we don't register which slit, we get the interference pattern. We also learn something new from this experiment: the counters behind slit 1 and slit 2 never both click simultaneously. This disposes of the crazy notion that a photon somehow splits into two pieces, and one piece goes through slit 2 while the other goes through slit 1. It also disposes of the other crazy notion that the photon checks whether slit 2 is open and then somehow doubles back to pass through slit 1. Nonetheless, we're left with deeply disturbing results: we now know that if we merely peek through which slit a given photon passes, we destroy the interference pattern; if we don't peek, the interference pattern comes back.

One final check we can do to this experiment is to make sure whomever or whatever is measuring, detecting, or peeking to see which slit the photon went through isn't somehow affecting where the photon appears on the detector screen by "biasing" the experiment. So we can place all the equipment in another room and record the events only on a disk drive and only read out the results for the final data later. We effectively make our experiment "double blind" to avoid any possible human bias of the outcome.

After repeating the experiment many times and comparing the data with the actions of the slit detectors, we find that whenever the slit detector in *on* and there is a record somewhere in the universe of which slit the photon went through, we do *not* see the interference pattern. Whenever the slit detector is *off* and we have no record of which slit a photon went through, then we *do* see the wavelike interference pattern. This is spookier still—the modest little photons seem to know if and when someone or something is watching them. They behave like particles, going through only one slit in a particle-like manner when they are observed as to which slit they went through. But they behave like waves, interfering as if having gone through both slits when no one or nothing is watching which slit they went through. Perhaps we all need a stiff drink at this point.

The double-slit experiment seems to be the ultimate shoot-out between the particle and wave concepts. It has revealed a shocking behavior of particles and the outcome of an experiment to our own interactions with it. Photons are indeed strange things, but it only gets more bizarre later when we redo the experiment with electrons—and *we find the same behavior*!

In Young's experiment, we understood the dark interference fringes as the result of a cancellation of two waves arriving at the detector screen from a combination of waves passing through both slit 1 and slit 2. This exact cancellation occurred at places where the travel distances from the two slits were just right to allow crests and troughs of waves to coincide. Now we've done the experiment with single photons, one at a time, and we got the same result. In fact, we must conclude that the light waves, discovered by Dr. Young and his followers, are real, but they are also really streams of particles called photons. Or, in reverse, we conclude that light is indeed composed of the corpuscles that Newton conceived of, but these corpuscles (photons) really behave like waves. It is neither. And it is both. Such is the mental torment of quantum physics.[14]

THROUGH A GLASS, BRIGHTLY

The photon as a particle simply "is." It clicks. It collides. The photon explains the emission of colors from the blackbody radiation; it explains the photoelectric effect and the Compton effect. It is! But it still cannot explain interference, and there is something else it cannot explain.

Recall our gazing through the store window of Victoria's Secret, our experiment from chapter 1. Let's take another stroll, past Macy's this time, where the spring fashions are on view in the large display windows, illuminated by the store lights and bright sunlight. You see the mannequins in their finery, but in the glass you also see a dim reflection of the street outside, including yourself. Let's say there happens to be a mirror in the display window. You see your image very clearly, very sharply in the mirror, but also much more dimly in the glass of the store window.

So we would reason as follows: sunlight reflected from you is transmitted through the window glass, hits the mirror, and reflects back to your eyeball. But a small fraction of the light is also reflected from the store window. So what, you may ask? This makes perfect sense anyway you look at it. If light is a wave, there's no problem. Part of a wave can be transmitted and part reflected as waves are wont to do. And if light is a stream of particles, some fraction, say, 96 percent of the particles, are transmitted through the glass and 4 percent bounce back. But if light is a stream of photons—lots and lots of photons, all identical—then how does a given photon (call him Bernie) decide whether to be transmitted or reflected?

Picture this: hordes of identical photons head for the glass window. Most pass through, but every once in awhile one photon bounces back. Remember that photons are considered particles—indivisible, uncuttable, irreducible—and no one has ever seen 4 percent or 96 percent of one photon. Bernie either gets through 100 percent of the time or is reflected 100 percent of the time. Perhaps there are lots of atoms in the glass and Bernie is simply hitting one of those 4 percent of the time and being reflected. But then we wouldn't see a nice mirrorlike dim image of reflection. We would see a slightly foggy glass, as 4 percent of the photons were lost. The reflection of the image of "us" is a nice coherent wavelike effect, yet it seems to work for individual photons as well. So we have a problem with partial reflections of individual particles. There seems to be a 4 percent probability of photons, as particles, ending up in a reflected wave as 96 percent of them end up in a transmitted wave. Einstein, in 1901, already saw that Planck's photon model for light would introduce this probabilistic element into physics, and he didn't like it. As time passed, he grew to hate it more and more.

THE WALRUS AND THE PLUM PUDDING

Meanwhile, as if Planck's solution of the ultraviolet catastrophe and Einstein's explanation of the photoelectric effect weren't enough, classical physics suffered a third major wake-up call in the early years of the twentieth century. We will call it the "failure of the plum pudding model."

Ernest Rutherford (1871–1937), a large, gruff walrus of a man who had already won a Nobel Prize for his work in radioactivity, was now installed as director of the famous Cavendish Laboratory in Cambridge, England.[15] Growing up as one of a dozen children in a hardscrabble farming family in New Zealand, Rutherford was weaned on hard work, thrift, and technological innovation. As a child he tinkered with clocks and made models of his father's waterwheels. By the time he was a graduate student, he was investigating electromagnetism and managed to devise a detector of wireless signals before Guglielmo Marconi began his famous experiments. When a scholarship brought him to the Cavendish Laboratory, he lugged his wireless device with him to England and was soon receiving signals from half a mile away, a feat that impressed a number of Cambridge dons, including J. J. Thomson, the Cavendish director at the time.

After the discovery of x-rays, which at that time were known as "Becquerel rays," Thomson invited Rutherford to join him in researching the effect of these "rays" on the discharge of electricity in gases. Though the New Zealander was pining for his native land, it was an offer he could not refuse. The fruit of this collaboration was a famous joint paper on ionization, the basic idea of which was that x-rays colliding with matter seemed to create an equal number of positive and negative carriers of electricity, or "ions." Later Thompson would declare, "I have never had a student with more enthusiasm or ability for original research than Mr. Rutherford."

By 1909, Rutherford's postdocs were shooting something called alpha particles at a piece of thin gold foil and mapping the way the particles were slightly deflected by the heavy gold atoms in the foil. Then something utterly unexpected happened. While most of the particles were deflected only slightly by their passage through the gold foil en route to a distant detector screen, one in eight thousand alpha particles bounced back toward the source. As Rutherford remembered it, "It was as if you fired a fifteen-inch artillery shell at a piece of tissue paper and it came back and hit you." What was happening? What kind of thing inside the atom was repelling the positively charged and massive alpha particle?

Thanks to the earlier work of J. J. Thomson, atoms were known to be full of very lightweight, negatively charged electrons. Obviously, for the atom to be stable it had to have an equal amount of positive charge, balancing the negative charge of the electrons. But where this charge resided within the atom remained a mystery. No one, before Rutherford, had any way of mapping the shape of the atom.

In 1905, J. J. Thomson had proposed a model of the atom in which the positive charge is dispersed evenly throughout a sphere covering the entire atom, with electrons embedded in it like plums in a pudding—dubbed by the physics community the "plum pudding model." Accordingly, in this model the alpha particles of Rutherford should always have charged straight through the atom—always! The atom was like a big glob of shaving cream, and the alpha particles were rifle bullets. Rifle bullets should tear straight through blobs of shaving cream. Imagine seeing a rifle bullet occasionally deflected and ricocheting backward upon colliding with a blob of shaving cream. Such was the observation of Rutherford.

According to Rutherford's calculations, though, there was only one way that alpha particles could ever be deflected backward, and that was if the entire mass and positive charge of the atom was concentrated in a "nucleus," a small volume in the center of the atom. The nucleus's hefty mass and large positive charge could repel the positively charged alpha particles that came within range. It was as if within the glob of shaving cream there were dense hard ball bearings that could cause the bullets to collide and deflect. The electrons were orbiting this dense central charge of the atom. Thus the pastry puff picture of the atom of J. J. Thompson was now in the trash bin. An atom resembled more a tiny solar system with miniature planets (electrons) orbiting a dense, dark star at the center (nucleus) that was held together by electromagnetism.

Further experiments indicated that the nucleus was indeed tiny—one-trillionth of the volume of the atom—even though more than 99.98 percent of the mass of the atom resided in the nucleus. But most of the atom was just *empty space*, with electrons interspersed and rapidly moving around within it. How amazing: *matter is mostly empty space—a void!* (The "solid" chair you're sitting on is overwhelmingly composed of nothing.) At the time of this discovery, within this tiny solar system model of an atom, all the laws of Newton and Maxwell—such as $F = ma$—were still thought to be rock solid, just as in the macroscopic solar system with the sun and its planets. The same laws of classical physics were believed to work in the atom just as they did everywhere else. Everybody slept well at night—until Niels Bohr showed up.

THE MELANCHOLY DANE

Niels Bohr, a young theoretical physicist from Denmark who was studying at the Cavendish Laboratory, attended a lecture by Rutherford and was so captivated by his atomic theory that he arranged to visit the great experimentalist for four months in 1912.[16] At that time Rutherford was working in Manchester.

Sitting down and thinking about the new data, Bohr quickly perceived something significant about Rutherford's model. It was a disaster! Applying Maxwell's equations to the electrons in their circular orbits around the nucleus, Bohr realized that, in their state of rapid circular motion, electrons would radiate all their energy away in the form of electromagnetic waves, very quickly. The orbits would quickly shrink to zero, within a tenth of a millionth of a billionth of a second, and the electrons would spiral down into the nucleus. This would make the atom, ergo all of matter, unstable and the physical world as we know it impossible. Maxwell's equations spelled disaster for a classical (Newtonian) atom. Either the model had to be wrong or the venerable laws of classical physics had to be wrong.

Bohr applied himself to understanding the simplest atom—the hydrogen atom—which has a single electron in orbit around a positively charged nucleus, à la Rutherford. Thinking about waves versus particles, the Planckian and Einsteinian ideas that were in the air, concerning wavelike aspects of the motion of particles that might be trapped in an atomic orbit, Bohr was led to propose a very anticlassical (and outrageous!) idea. Bohr argued that only certain special orbits can happen for electrons in atoms because the motion of electrons in these orbits is like that of waves. One of these special orbits would be the one with the least amount of energy, where the electron is moving closest to the nucleus. In this orbit the electron cannot radiate away any more energy. Since this is the state of lowest possible energy for the electron, it has no lower energy state into which to go. This special orbit is called the *ground state.*

One of the key facts that Bohr was trying to account for was the discrete spectral lines of emitted and absorbed light by atoms, as we've discussed previously. Recall that when various elements are heated until they glow, each element when viewed through a spectrometer is seen to emit a distinctive series of sharp, brightly colored lines of light superimposed on a darker glow of continuous colors. At the same time, the sun's spectrum was found to be overlaid by a series of fine dark lines at particular places. The bright lines were found to represent emissions—the dark lines, absorption. Like other elements, glowing hydrogen emits a series of spectral lines that are like a fingerprint, and these were the experimental data that Bohr attempted to explain with his fledgling model.

In three papers published in 1913, Bohr articulated his audacious quantum theory of the hydrogen atom. Each of the atom's magic orbits is characterized by a certain energy. An electron emits radiation when it "jumps" from an orbit of higher energy, say E_3, down to one of lower energy, say, E_2 . It emits a photon whose energy ($E = hf$) is given by the difference of the energy of the two orbits: $E_3 - E_2 = hf$. With billions of atoms doing this at the same time, we see a bright spectral line. In a model that preserved some of Newton's mechanics but disposed of what didn't give the right answers, Bohr triumphantly calculated the wavelengths of all the spectral lines contained in glowing hydrogen. His formula gave these in terms of known quantities such as the charge and mass of the electron (embellished by assorted things like 2, π , and, of course, the quantum logo of Planck, h).

Thus, in Bohr's quantum picture, the electron must limit itself to specific, seemingly "magic" orbits, corresponding to well-demarcated energy states of the atom. The levels are numbered 1, 2, 3, 4 . . . , and each one has an energy E_1, E_2, E_3, E_4, . . . , and so on. An electron can only absorb energy in "bundles," or quanta. If the electron swallows the right quanta, it can "jump" to a higher energy level, for example, from E_2 to E_3. And the electrons in higher energy states spontaneously tumble down to the lower states—from E_3 back down to E_2 and so on, emitting photons, or quanta, of light. These quanta can be observed as specific wavelengths, the spectral lines, predicted exactly for hydrogen in Bohr's model.

THE CHARACTER OF THE ATOM.

Thanks to Rutherford and Bohr, we have those iconic drawings of the atom for the logo of the Atomic Energy Commission, with the nifty electrons whizzing around like tiny planets in elliptical Keplerian orbits. Many people today probably believe that is how the atom really looks. It doesn't, alas, because Bohr, though inspired, wasn't exactly right. His dazzling success turned out to be a bit premature. Bohr could explain some aspects of the hydrogen atom—the simplest ones—but he couldn't explain helium, the next-simplest atom with two electrons. As the 1920s approached, scientists still had not formulated a proper theory for quantum mechanics. What we had was a first step, called "Bohr's old quantum theory."

The founding fathers—Planck, Einstein, Rutherford, and Bohr—launched a revolution yet to reach fruition. Obviously, we were not in Kansas anymore, as we were now dealing with quantum leaps, electrons mysteriously jumping from one magic orbit to another without ever being in between, and photons that were waves and particles, neither or both at the same time. Much still remained to be understood.

Out of the mid-wood's twilight
Into the meadow's dawn,
Ivory limbed and brown-eyed,
Flashes my Faun!

He skips through the copses singing,
And his shadow dances along,
And I know not which I should follow,
Shadow or song!

O Hunter, snare me his shadow!
O Nightingale, catch me his strain!
Else moonstruck with music and madness
I track him in vain!

Oscar Wilde, "In the Forest"[17]

Chapter 5

HEISENBERG'S UNCERTAINTY

Here is the moment you've been waiting for. We now head straight into quantum mechanics proper, contemplating a terrain so mystifying and often alien that it inspired one of physics' greatest luminaries, Wolfgang Pauli, to seriously consider quitting in 1925. "For me," he wrote in exasperation to a colleague, "physics is too difficult and I wish that I were a film comedian or something similar and had never heard of physics." If the formidable Pauli had dumped physics to become the Jerry Lewis of his generation, we might never have had the "Pauli exclusion principle," and the history of science might have taken a markedly different turn.[1] But, fortunately, he stuck it out—as we hope you will, too. Although the journey is not for the faint of heart, it will end up being tremendously rewarding.

NATURE IS LUMPY

Let's begin with the quantum theory that Niels Bohr originally formulated as a result of Rutherford's experiment. It revealed that atoms were not made of plum pudding but rather had a dense core at the center surrounded by electrons whizzing about, something like a solar system with a sun at the center and little planets orbiting around it. As we mentioned, Bohr's original "old quantum theory" eventually died and went to theory heaven. As the quantum theory was refined, the old quantum theory of Bohr, with its crazy stew of classical mechanics and ad hoc quantum rules, was eventually discarded. Nonetheless, Bohr introduced the world to the quantum atom, and some of the theory's original implications gained credibility from the results of a brilliant experiment.

Under the laws of classical physics, an electron could not orbit the nucleus of the atom. When placed in orbits, electrons must accelerate—in fact, all circular motion is accelerated motion, since the velocity is continuously changing direction with time. And, according to Maxwell's theory of electromagnetism, accelerated charges must radiate energy in the form of electromagnetic radiation—that is, light. Estimates showed that *all the electron's orbital energy would be almost instantaneously radiated away* into electromagnetic waves, and the electron would spiral down, like a wounded bird, crashing into the nucleus. The orbits of the electrons, and the atom itself, would thus collapse. Such collapsed atoms would be chemically dead and useless. Nothing about the energy of electrons, atoms, or nuclei seemed to make sense in the classical theory. It demanded the invention of some new description: the quantum theory.

Moreover, scientists of the late nineteenth century knew that atoms do emit light, but only in distinct spectral lines having definite colors, or distinct, or *quantized* values of the wavelength (or frequency). It seemed as if only certain special electron orbits existed in the atom, the electron hopping to and fro between these orbits as it emitted or absorbed light. A Kepler-like picture of the orbits would have predicted a continuous spectrum of radiated light, since there are a continuous set of possible Keplerian orbits. It was as if the world of the atom was "digital" and far from the continuously varying world of Newtonian physics.

Bohr focused on the simplest atom, hydrogen, which has a single negatively charged electron orbiting a heavy nucleus (a proton). He played around with the new ideas of quantum theory, à la Planck and Einstein. Bohr ultimately noticed that Planck's idea of associating a certain wavelength (or frequency) with the momentum (or energy), if applied to electrons, might imply the existence of certain special orbits, and he finally hit upon the formula for the energies of the electron orbits. Bohr's special orbits were circular and each had a certain circumference (distance around the orbit in one period). And Bohr insisted the circumference should always equal the quantum wavelength of the electron, derived from Planck's formula.[2] Each of these magic orbits corresponded to a particular energy so that atoms would possess a series of discrete energy states.

Immediately Bohr realized that there was a smallest orbit for the electron, where the orbit circumference was the smallest and the electron was closest to the nucleus. Once the electron fell into this orbit, no longer could the fateful dive of the electron into the nucleus ever occur. This smallest orbit is called the "ground state," the state of lowest energy, and the electron cannot go into any state of lower energy, thus stabilizing the atom. A ground state is a feature of all quantum systems. The vacuum is the ground state of our entire universe.

The new ideas worked too well. Out popped important numbers that characterized the observed radiation patterns seen in experiments for hydrogen. All the electrons in atoms are in, what physicists call "bound states." Without adding more energy, the electrons always remain circling the nucleus in their orbits. The amount of energy you must add to pull the electron away from the atom and set it free is called the "binding energy." And the binding energy depends on which orbit the electron is in. We usually define the energy of a free, uncaptured electron of zero velocity to have zero energy of motion (this is actually arbitrary; we could add any value to this energy we like, but it is convenient to define this as the state zero energy); then the *binding energy* of a bound electron would be given by a negative number, since the bound state has less energy than the lowest free state. Likewise, if a lowly free electron gets captured into an orbit in the atom, the amount of energy it radiates away into light during the capture process is equal to the (magnitude of the) binding energy of the orbit into which it is captured.

The binding energies of the Bohr orbit states are measured in a unit called the "electron volts" (eV).[3] The ground state, that special orbit closest to the nucleus, has a binding energy of 13.6 eV (this means it takes 13.6 electron volts of energy *to remove* an electron from the ground state of a hydrogen atom). This number, 13.6 eV, is often called the *Rydberg*, after the Swedish physicist Johannes Rydberg who in 1888 had guessed (with Johann Balmer and others) a formula for the spectral lines of hydrogen and other atoms. So this special number and the formula that predicted the binding energies was well known for many years prior to Bohr, but with Bohr there was now a formula that logically explained what was going on.

The quantum states of an electron in hydrogen (equivalent to Bohr's orbits) can be represented symbolically as a series of numbers: $n = 1, 2, 3, \ldots$.

The state with the largest magnitude (most negative) binding energy—the ground state—would correspond to $n = 1$; the first excited state would be $n = 2$, and so on. That these discrete states or conditions are the only allowed conditions for an atom is the essence of quantum theory. The integer n is dignified by a name: "the principal quantum number." Each state, or quantum number, is characterized by a quantity of energy (the eV numbers above) and is labeled E_1, E_2, E_3, and so on (see note 3).

Recall, too, that in this bygone, but not forgotten theory, atoms can radiate a photon by jumping from a state of higher energy to a lower energy. This rule, of course, does not apply to the E_1 energy level, or $n = 1$ electron—the electron in the ground state—as it obviously has nowhere to jump down to. These changes in state, called "transitions," operate in a predictable, mathematical manner: Say the $n = 3$ electron jumps down to the $n = 2$ state, then $n = 2$ jumps down to $n = 1$. For each jump the electron must then emit a photon whose energy would be equal to the difference in energies between the two atomic states, as in $E_3 - E_2$, or $E_2 - E_1$. Expressed as the energy states listed above, that might be 10.5 eV – 9.2 eV = 1.3 eV, or 13.6 eV – 10.5 eV = 3.1 eV, these being the energies of the observable emitted photons. Because the energy and wavelength (λ, pronounced LAM-dah) of a particle of light (a "photon") are related by Planck's formula, $E = hf = hc/\lambda$, physicists could determine the energies of the electron's states by using a spectrometer to measure the wavelengths of the light emitted from atoms. This picture worked beautifully for the spectral lines emitted by the simple atom of hydrogen (one electron around a nucleus), but it wasn't clear what to do for helium, the next-simplest atom.

Bohr suggested that the states of motion of electrons could also be measured in quite a different way: by letting atoms absorb energy. If the states were real, the absorption would be in the form of lumps. Energy could only be absorbed if it were exactly equal to the upward jump from E_1 to E_2, or E_3, and so on. The critical experiment to test the absorption idea was carried out in 1914 by James Franck and Gustav Hertz in Berlin, in what was probably the last significant experimental research in Germany before World War I. Although their data perfectly matched Bohr's analysis of the emission process, the German experimenters knew nothing about the great Dane's theory until much later.

THE FRANCK-HERTZ EXPERIMENT

Before we describe the experimental details, let's visualize a very crude classical analogy. Imagine rolling some small steel balls down a hill. At the bottom of the hill is a slight rise, requiring the balls to have sufficient energy to climb a bit before falling into a bucket. Now let's drive some steel pegs into the hill at random locations so that the surface somewhat resembles a pinball machine. Our steel balls bounce off the pegs on the way down, but since the collisions of the balls with the pegs are *elastic*—no energy is lost as steel recoils against steel—the balls still have enough energy to make it up the speed bump at the bottom and then plop into the bucket. If we replaced the steel pegs with pegs made of Silly Putty, however, the collisions would be *inelastic* (the putty would absorb energy), and the balls, now much reduced in energy, would dribble aimlessly to the bottom, incapable of making it over the rise. Now assume that we can adjust the height of our hill to give our balls more or less energy as they arrive at the bottom.

Franck and Hertz did something similar, except they used electrons, instead of our steel balls above, produced from a heated filament. These electrons were attracted to a wire screen through a low-pressure gas of mercury atoms, the equivalent of our metal plugs. The screen's voltage, which could be adjusted from zero to thirty volts, acted as the hill. That is, the screen attracted the electrons, giving them energy, just as the descending slope supplied energy to our steel balls. After the electrons bounced around through the mercury atoms and made it to the screen, they got hit with a "retarding voltage" of one volt, the equivalent of our little speed bump at the bottom of the hill. If the electrons overcame the retarding voltage (climbed the small rise), they then hit a collection plate (our bucket), where their current was recorded. The game was to measure this current as the screen voltage was slowly increased. The rising voltage effectively increased the violence of the collisions the electrons made with mercury atoms.

The crucial data are presented in a graph in which the resulting current, I, is plotted against the voltage, V. The key idea is this: if a collision with the mercury atoms resulted in a loss of energy (inelastic collisions) and if this took place near the screen, then the electrons could not overcome the retarding "rise" and would fail to be collected. If the collision did not result in a loss of energy (an elastic collision), then the electrons would arrive at the screen with the full energy (determined by V) and would sail up the incline to contribute to the current.

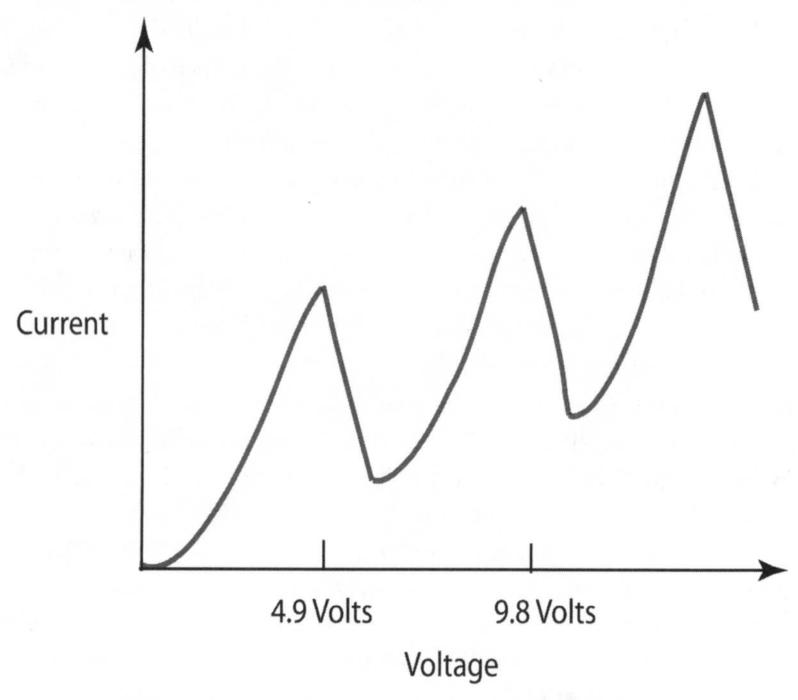

FIGURE 21: The Franck-Hertz experiment. As we increase the voltage (which delivers energy to electrons comprising the current) in a gas of sodium vapor, the current increases until the energy necessary to excite the sodium atoms is reached, for example, at 4.9 volts. The sodium atoms absorb the energy through collisions with the electron current and jump to their next energy level, depleting the electron current. The light of the sodium atoms decaying back down to the ground state can be detected. As the voltage is increased to 9.8 volts the second transition energy is reached, and again the atoms absorb energy depleting the current by collisions. The Franck-Hertz experiment confirms Bohr's prediction based on his theory of the atom.

What one sees, as we start with low voltage, is that the current increases as soon as V exceeds the retarding voltage. This rise in current indicates that the many collisions of electrons with mercury atoms do not result in any loss of energy. However, something strange happens when the voltage reaches a critical value of exactly 4.9 volts. There we see an abrupt drop in current. Evidently, when electrons reach an energy of 4.9 eV, the collisions with the mercury atoms do result in a loss of energy. The electrons lose energy and cannot make it up the "hill" to the collector.[4]

Bohr delightedly explained it all: atoms can only absorb energy in packages that correspond to the allowed energy states. The energy gap between the ground state of mercury, E_1, and its first excited state, E_2, is 4.9 eV. An electron of this energy can give all of its energy to the mercury atom, recoiling with zero energy. Then it could not exceed the retarding voltage and could not be collected. Conversely, an electron with 4.6, 4.7, or 4.8 eV, and so on, is close but no cigar: it would give none of its energy to the mercury atom and recoil elastically, sailing through the retarding screen, and would be measured as current. However, as the voltage V would continue to rise, the electron would reach its critical energy of 4.9 eV some distance away from the screen and, recoiling from its inelastic collision, could yet gain enough energy to overcome the retarding voltage and add to the current (get "collected"). So the curve would resume its rise. At 9.8 volts, though, what would happen? Another sharp drop occurs as the electrons would experience two inelastic collisions before reaching the screen, raising two mercury atoms to their E_2 state while the electrons would deplete their energy.

Fascinating, but is this positive proof of Bohr's hypothesis? Let's see: The excited mercury atoms don't stay excited. After a very brief interval they would "de-excite," that is, "jump" back down to the ground state, emitting a photon in the process. The photon's wavelength would be determined by the change in energy between the two states, or exactly 4.9 eV for the first peak in the curve. This wavelength is in the purple color range, the characteristic light emitted by a mercury arc lamp. Aiming a spectrometer at the gas, Franck and Hertz looked for the purple line. At a voltage of less than 4.9 V, no light was seen. But at precisely 4.9 V, the line appeared! What they were seeing was the de-excitation of mercury atoms that had been excited by collisions with electrons, which had stolen all the energy away from the electrons.

Thus, discrete energy levels in atoms are real. The classical belief in an intrinsic continuity of nature now became dead. This experiment made it into the history books, and it was called the "Franck-Hertz experiment."

THE TERRIBLE TWENTIES

It is hard to appreciate the state of panic that gripped the world's leading physicists at the beginning of the Terrible Twenties, in the years 1920 to 1925. After four hundred years of trusting that nature followed a classical rational plan, scientists suddenly were forced to reconsider this core belief. What shattered the comfortable old worldview was, first and foremost, a profoundly unsettling duality in the quantum world. On the one hand, there had been a series of experiments, repeated many times, demonstrating that light was a wave, exhibiting the interference and diffraction phenomena typical of waves. We have already seen in graphic detail how it is impossible to explain the double-slit experiment if light consists of particles.

At the same time, equally compelling data shouted out that light consisted of particles. As we have already observed in chapter 4, the study of blackbody radiation, the photoelectric effect, and Compton's experiment with electron-photon collisions all revealed this irreducible "particleness." The logical conclusion of this group of experiments could only be that light of a given color, and therefore of a given wavelength, was a stream of particles. Each of these particles moved at the velocity of light, *c*, and each had a certain momentum. Momentum—in Newtonian terms the product of velocity and mass—is a significant characteristic of the motion of matter, which any highway patrolman can tell you. For photons, the momentum is the energy divided by the velocity of light, *c*. Momentum is an important concept because the total momentum of all the objects about to collide is conserved, that is, it is the same after the collision and it never changes throughout the process. For instance, when two billiard balls collide, the sums of the masses times the velocities before the collision will equal the sums of masses times velocities after the collision.[5] Compton's experiment had shown that light quanta, like automobiles and other large objects, also conserve the total momentum in collisions.

Now let's take a moment to clarify the difference between particles and waves. First, particles possess the quality of discreteness. Take a glass of water and a glass of fine, dry sand. Both can be poured, both swirled, and—if one doesn't look too closely—their properties are quite similar. But while the liquid appears continuous and smooth, the sand consists of countable, discrete grains. A small scoop will always pick up a volume of smooth liquid, but only one, two, three, . . . , a countable number of grains of sand. In quantum theory, primary integers become crucial—a flashback to Pythagoras, the ancient Greek mathematician. A particle has a well-defined location in space at any time and moves on a trajectory unlike a wave, which is spread over space. And particles have energy and momentum, which can be transferred to other particles in collisions. By definition, to be a particle is to not be a wave, and vice versa.

Now back to our story. To their befuddlement, physicists were encountering an outlandish beast—a wave particle, even a *wavicle*, as some dubbed it. Although light was well known to consist of waves, experiment after experiment revealed photons (light quanta) as tiny lumps capable of colliding and pushing electrons around. They could be absorbed by matter, either swallowed whole or not at all. Excited atoms could emit these photons, losing the exact amount of energy, $E = hf$, that the photon carried off. This state of affairs gained a further new dimension when a French physics student, a young aristocrat named Louis-Cesar-Victor-Maurice de Broglie, wrote an amazing doctoral thesis.[6]

Scandalized that one of their own was contemplating a career in dowdy physics instead of the military, diplomacy, or politics, de Broglie's family had at first resisted Louis's aspirations. His grandfather, the duke, had scoffed, "science is an old lady content with the attractions of old men." So the young de Broglie was forced to compromise, pursuing (for a time) a career in the navy and in his spare time experimenting in his own laboratory that was set up in the familial mansion. In the navy he made a name for himself in wireless work, but after the old duke's death he was permitted to retire and dedicate himself full-time to his true passion.

De Broglie had been meditating on Einstein's anxiety about the photoelectric effect, as well as his proof of the photon nature of light and its incompatibility with light's well-established wave properties. Rereading Einstein, de Broglie was seized with a most unorthodox notion. If, he reasoned, light waves seem to have particle properties, perhaps the reverse is also true. Perhaps particles—all particles—themselves demonstrate wave properties, or, as Louis put it (of Bohr's theory of the atom), "this fact suggested to me the idea that electrons too could not be considered simply as particles, but that frequency (wave properties) must be assigned to them also."[7]

Normally, so brash a doctoral topic would have earned the student a transfer to the department of theology or to a junior college in Lower Slobovenia, but this was 1924, and de Broglie had an influential cheering section. The great Albert Einstein, who was called in to vet the candidate's paper when the University of Paris examiners conceded their perplexity, expressed great interest in de Broglie's idea (we wonder if Einstein thought, "I should have thought of that!") and incorporated it in his quantum studies. The Master wrote back to the Paris dissertation committee that "de Broglie has lifted a corner of the great veil." Not only did de Broglie get his PhD, but shortly thereafter he earned a Nobel Prize for his doctoral dissertation. Mainly, he precisely connected the Newtonian momentum of an electron (mass times velocity) to the wavelength of the "electron wave" through Planck's formula.[8] But *electron waves*? Electrons are particles. Where is the wave? De Broglie equivocated, citing "some mysterious internal periodic process" going on inside the particle. This sounds vague, but it's exactly what he intended. Despite his vagueness, though, de Broglie was on to something.

At the time, in 1927, two American physicists at AT&T's renowned Bell Labs, in New Jersey, were engaged in studying the properties of vacuum tubes by shooting streams of electrons at various oxide-coated metal surfaces. The electrons created strange patterns in the way they emerged from these crystals: in some directions many electrons streamed out of the crystal, while in other directions no electrons were detected. This had puzzled Bell Labs' physicists until they learned of de Broglie's crazy electron wave idea. The crystal was just a complex version of the two-slit experiment of Thomas Young. The electrons were demonstrating the normal property of wave interference, known as diffraction! The patterns made sense if the wavelength of the electron waves was indeed related to the electrons' momentum, precisely as predicted by de Broglie. The regular spacing of atoms in the crystal served as the equivalent of the "slits" in the famous double-slit experiment of Young, some two hundred years earlier. These crucial "electron diffraction" experiments verified the de Broglie connection between momentum and wavelength. Electrons are particles that behave like waves, which was fairly easy to see.

We'll return to the subject of diffraction a bit later, shooting electrons through our familiar double-slit experiments—and we'll find an even more shocking result. Moreover, it is the diffractive motion of electrons, as waves, in a crystal that gives rise to the behavior of various materials as good conductors of electricity, or insulators, or semiconductors. Ultimately it has yielded such devices as the transistor. But before we get there we must introduce another of the heroes—perhaps the superhero—of the quantum revolution.

A STRANGE MATHEMATICS

Werner Heisenberg (1901–1976) was a theorist's theorist, nearly flunking his oral qualifying exam at the University of Munich because he had no clue how a battery worked. Fortunately for what would become the field of quantum physics, he squeaked by. Still, Heisenberg had many other things going for him. During World War I, while his father was off fighting as a reserve infantry man, food and fuel shortages often forced the university to close.

In the summer of 1918 the young Werner, weakened and half starved, was forced to educate himself for a time, though he brought in the harvest of a Bavarian farm along with other local schoolboys. In the 1920s he was a twenty-three-year-old wunderkind, a near-concert-level pianist, a consummate hiker and skier, a classical scholar, as well as a mathematician-turned-physicist. As a student of the eminent elder-day physicist Arnold Sommerfeld, he met fellow student Wolfgang Pauli, who was to become his closest collaborator and sharpest critic. In 1922 Sommerfeld took young Heisenberg on a field trip to Göttingen, the great intellectual center of Europe at that time, for a series of lectures on the newborn quantum atomic physics by Niels Bohr. It was at these lectures that the youthful Heisenberg, no shrinking violet, had the effrontery to criticize some of the great man's assertions and to question his core theoretical model of the atom. Nonetheless, this confrontation marked the beginning of a lifelong collaboration and mutual admiration.[9]

From that day on, Heisenberg got drawn deeply into the quantum dilemma. He took a semester in Copenhagen in 1924, working with Bohr on problems related to the absorption and emission of radiation, where he gained respect for Bohr's "philosophical thinking" (in Pauli's words).[10] As he struggled to visualize the reality of the Bohr atom, with its whimsical planetary electron orbits, Pauli became convinced that there was something wrong with the picture. The more he pondered the image, the more he suspected that the neat, almost circular Bohr orbits were merely intellectual constructs and extra baggage, so he developed the notion that the very idea of electrons occupying orbits was a hang-over from the classical world of Newton.

The young Werner developed a ruthless credo: no mental pictures based on classical thinking, in other words, no tiny solar systems, even if they make good beer ads. The road to salvation is not good graphics but sharp mathematical reasoning. Furthermore, he held that one should ruthlessly dispense with any concept (such as orbits) that cannot be directly measured.

What could be measured about the atom were the distinct spectral lines, the emission or absorption of light by atoms that results from the "quantum jumps" made by the electrons changing their orbits in atoms. So it was to the spectral lines—the visible, testable clues to the hidden and unknowable motions of electrons—that Heisenberg turned his attention. This was the fiercely difficult problem that the hay fever–afflicted Heisenberg took with him to the North Sea island of Helgoland in 1925.[11]

Guiding his thinking was a principle of Bohr, the "principle of correspondence," which preached that quantum laws should be so constructed as to blend seamlessly with classical laws when the systems described get large enough. But how large? The answer is: large enough so that Planck's constant, h, becomes a negligibly small factor in the relevant equations (e.g., the equations of a rocket launched into space do not involve h because all the ingredients, rocket engine, fuel, astronaut, are macroscopic). Whereas an atomic object might have a mass of 10^{-27} kilogram, the mass of a barely visible speck of dust might be 10^{-7} kilogram—yet that speck is heavier than the atom by a factor of 100,000,000,000,000,000,000! (A one followed by 20 zeroes, as noted earlier, is more easily written as 10^{20}.) This places the lowly dust speck solidly in the classical regime; it is a macro object, and its physical behavior is unaffected by Planck's constant. The more fundamental quantum laws naturally account for atomic-sized phenomena, but as they are applied to bigger, aggregate, macro-phenomena, the details of quantum atoms drop out and the description then blends with Newton's laws and Maxwell's equations. The key to "correspondence" (which we emphasize here and elsewhere) is that the brand–new, weird, unfamiliar quantum concepts should, as objects get bigger and bigger, "correspond" directly to classical concepts in the macroworld.

Guided by Bohr's correspondence principle, Heisenberg introduced the familiar and mundane classical quantities, such as position, velocity, and acceleration, to describe the electron, in order for it to have any correspondence with the Newtonian world. But he found that his efforts to reconcile the quantum and classical realms required that a strange new "algebra" be introduced into physics.

Every schoolchild learns that when we multiply two numbers, such as a times b ($a \times b$), it is the same as b times a ($b \times a$), that is, $a \times b = b \times a$. For example, $3 \times 4 = 4 \times 3 = 12$. This is called the *commutative law of multiplication*. However, there existed already in the minds and literature of mathematicians certain purely mathematical systems of numbers that are noncommutative, systems where $a \times b$ does not equal $b \times a$. It's not hard to see that such things exist in nature (a book rotated through two successive rotations, in two different orders, does not satisfy the commutative law).[12]

Heisenberg was not schooled in the pure mathematics of the era, but his more mathematically erudite colleagues quickly recognized this as the well-known algebra of matrices with complex numbers, that is, "matrix algebra," an exotic, sixty-year-old system that spelled out procedures for multiplying and adding arrays of numbers (matrices). Putting it all together, Heisenberg's new formalism yielded the first concrete proposal for what quantum physics is. He obtained sensible real numbers for the energies of the atomic states and the atomic transitions for the emission of light of an atom when an electron jumps from one state to another.

Moreover, when the new matrix algebra was applied to the hydrogen atom (one proton for the nucleus, and one "orbiting" electron) and other simple atomic systems, it worked beautifully. The solutions agreed with the experimental results. But now another profound insight leaped out of the arcane formulas of matrix mechanics.

THE INCEPTION OF THE UNCERTAINTY PRINCIPLE

The heart of *non-commutativity* of x and p is that, for a particle, both the position (say, in the x-direction) and the momentum (also in the x-direction) *cannot be measured to have definite values simultaneously*. In other words, if you measure the position exactly, you will necessarily disturb in an unknowable way the momentum, and vice versa. It isn't an issue of the measuring apparatus, or an inept experimentalist, rather it is the basic quantum physics of nature.

The matrix mechanics allowed this to be phrased in a statement that has sent philosophers up the trees ever since: "The uncertainty in the position of a particle, which we call Δx ('delta x'), and the uncertainty in the momentum, which we call Δp ('delta p') are related as: $\Delta x \, \Delta p \geq \hbar/2$ ('the product of the uncertainty in position times the uncertainty in momentum is always greater than or equal to Planck's constant divided by four times pi') where $\hbar = h/2\pi$." This means that if we make the uncertainty in the position measurement, Δx, as small as possible, we will always make the uncertainty in the momentum Δp arbitrarily large. And vice versa. In short, you simply can't have it both ways: either you measure the position precisely and forsake knowledge of the momentum or you measure the momentum (velocity) precisely and forsake knowledge of the position of a particle.

From this we can begin to see why the atom of Bohr cannot collapse, that is, why it must have a ground state, unlike in the case in Newtonian physics. For the atom to collapse, the electron would have to spiral down, down, and down into the nucleus, becoming ever more well localized in its position, that is, the uncertainty in the position would become nearly zero, or $\Delta x = 0$. But by Heisenberg's uncertainty principle, this implies that the momentum uncertainty, Δp, would have to grow arbitrarily large, and this would make the energy arbitrarily increase.[13] There is therefore a state that strikes a balance, in which the electron is "sort of" localized around the nucleus, with a nonzero Δx, such that the energy is the least possible, given the uncertainty Δp in momentum.

The physical origin of the uncertainty principle is easier to understand when we take the next step, à la Schrödinger. It's a common nonquantum property of waves, well known to telecommunications engineers. All of the above is tipping us off to the fact that we are describing some kind of wave phenomenon in quantum physics. At first it certainly looked as if "Heisenberg's matrix mechanics" was the only way to penetrate the world of the atom. But fortunately, in 1926, just as physicists everywhere were honing their matrix skills, along came another, more picturesque solution.

THE LOVELIEST EQUATION EVER WRITTEN

We first heard about Erwin Schrödinger and his famous vacation back in chapter 1. What Schrödinger did on this vacation, above all, was to develop an equation that significantly clarified what the quantum theory was (we'll refer to it as the Schrödinger equation).

Why all the fuss over an equation? First consider Newton's "equation of motion," $F = ma$, which governs the motion of baseballs and all other macroobjects moving under the influence of forces. The equation states that, given an applied force, F, an object of mass m will accelerate (change its velocity with time) with acceleration, a, such that $F = ma$. The solution of this equation gives us the position of the baseball and its velocity at any given time. Most of the work is in knowing what F is, and then grinding out the position, x, and velocity, v, at some time, t, from the acceleration, a. The relationships between position and velocity and acceleration are defined by Newton's differential calculus and can be difficult to determine (e.g., when there are very many particles with their own positions, all to determine at once). The equation is simple looking enough, but its applications get quite complicated, and it need not always have simple solutions.

Newton astounded the world by showing that a gravitational force (as defined by his universal law of gravitation) and his equation of motion produce the neat elliptical orbits and laws of planetary motion that Kepler had determined for the solar system. The same equation describes the moon, an apple falling from a tree, and a spacecraft on its way out of the solar system. But it cannot be analytically solved when there are four or more particles moving and interacting simultaneously under gravity, without either making approximations and/or using computers. That's the issue—the simple equation is at the heart of nature, yet it reflects the incredible complexity of our world. Schrödinger's equation is the quantum version of

$F = ma$. The solutions of Schrödinger's equation, however, do not yield positions and velocities of particles in the manner of Newton.

When Schrödinger went on vacation in December 1925, he brought along, in addition to his accommodating mistress, de Broglie's doctoral thesis on particles and waves. Few people had paid much attention to de Broglie's ideas, but Schrödinger would change all that. By March 1926, this undistinguished fortyish University of Zurich physics professor, rather geriatric by the standards of theorists of the day, published a single equation explaining the behavior of electrons in terms of de Broglie's waves (so much more satisfying to imagine than the cold abstractions of matrices). The lead player in the Schrödinger equation is something called Ψ ("psi" pronounced "sigh"), which is called the "wave function." The wave function is what Schrödinger's equation yields as its solution.[14]

Long before the quantum theory was promulgated, physicists were quite used to describing (classical) waves in continuous material media, such as sound waves in air (containing many, many particles). Consider, for example, a sound wave. We describe this with a mathematical quantity representing the pressure of the air, Ψ. Mathematically, $\Psi(x, t)$ is a "function," that is, it specifies the increase in the pressure of air in the wave, relative to normal constant room pressure, at any point in space, x, and at any time, t. A "traveling wave" arises naturally—it is actually the solution to the equations that describe the motion of air (or water, or electric or magnetic fields, etc.) when it is disturbed. So, too, are breaking waves, or tsunamis, or any of the many forms and shapes of water waves. These are all described by "differential equations," equations that involve calculus and typically determine in a unified way how many different things evolve in time and space. A particular differential equation is known as the "wave-equation," and it determines the "wave function" of a disturbance, $\Psi(x, t)$, such as the sound pressure in a sound wave at any point in space, x, at time t.

Schrödinger, in reading de Broglie, very quickly had an insight and noticed that the daunting mathematical formalism of Heisenberg could be written in a way that made it look very similar to the familiar equations of physics that describe wave disturbances. Therefore, one could say that, at least formally, the correct description of a quantum particle involved a new mathematical function, $\Psi(x, t)$, which was what Schrödinger dubbed the "wave function." Using the machinery of the quantum theory as interpreted by Schrödinger, that is, by solving Schrödinger's equation, one can, in principle, compute the wave function for a particle in almost any circumstance. At that time, however, no one yet knew what the wave function of quantum theory represented.

In quantum mechanics, therefore, we can no longer say "at any given time, t, the particle is located at a position x." Rather, we say that "the quantum state of motion of the particle is the wave function, $\Psi(x, t)$, which is the *quantum amplitude* Ψ at the time t at a position x." The precise position of the particle is no longer known. Only if the wave amplitude is known, and if it is large at some particular position, x, and near zero everywhere else, then can we say that the particle is "located near that position." In general, the wave function may be spread out in space, like a traveling wave, and then we never know, even in principle, exactly where the particle is. Bear in mind, at this stage of the development, physicists, including Schrödinger, were still very unclear about what the wave function really was.

Here, however, comes a twist in the road that is a stunning aspect of the mathematics of quantum mechanics. Schrödinger found that the wave function describing a given particle is, like any wave, a continuous function of space and time, but it must take on numerical values that *are not ordinary real numbers*. This is very much *unlike* a water wave or an electromagnetic wave, which is always a *real number* at each point of space and time. In a water wave, for example, we can say that "the height of the waves from trough to crest is ten feet, so the amplitude is five feet, and we are issuing a small craft advisory." Or we say that "the amplitude at the beach of the approaching tsunami is fifty feet, so this is a huge wave! Run!" These are real numbers that can be measured with a variety of instruments, and we all understand what they mean.

The quantum wave function, however, has values for its amplitude that are things called *complex numbers*.[15] For a quantum wave, we would say that "over at this point in space, the quantum wave has an amplitude of $0.3+0.5i$," where $i=\sqrt{-1}$, that is, *i is the number which when multiplied by itself gives* −1. Numbers that are (real) plus (real) times (i) are called *complex numbers*. In fact, Schrödinger's wave equation itself always involves $i=\sqrt{-1}$ in a fundamental way, and this is what forces the wave function to be a complex number.[16]

This mathematical twist of requiring complex numbers on the road to the quantum theory is inescapable. It strongly hints that *we can never directly measure the wave function of a quantum mechanical particle*, since we can only measure, in experiments, things that are always real numbers. From Schrödinger's viewpoint, electrons actually were waves—matter waves—no different from sound waves, water waves, and so on. But how could this be? A particle—say, an electron—has a well-defined location; it isn't spread out all over space. But, if we superimpose many waves on top of one another, we can arrange the sum so we can get a robust result in one place in space and essentially a complete cancellation at all other places in space. Thus waves, artfully put together, can represent something very localized in space that we might be tempted to call a particle. The particle would spring into being wherever we have a big lump in the addition of many waves. In this sense, a particle is like a "rogue wave" on the ocean where many small waves pile up at one place and make a humongous wave capable of toppling a ship.

FOURIER SOUP (OR, I THINK WE'RE BACK IN KANSAS)

The idea that Schrödinger's equation, which generates waves as solutions, can somehow imitate a localized particle is worth examining a bit more. Waves are disturbances that wiggle up and down. A wavelike disturbance typically stretches out over long distances in space. A particle, on the other hand, is, by definition, located somewhere. How can the extended stretched-out waves equal a localized particle?

A Frenchman named Jean Baptiste Joseph Fourier was a mathematician in late eighteenth-century to early nineteenth-century France who devised the mathematical method of adding (superimposing) up a large number of waves so cleverly that their net disturbance could be localized to a small region of space (much like a particle).

Let's say we have thousands and thousands of harmonic sound waves, all of different wavelengths. Each one of the many waves is a solution to a wave equation. The sum of a large number of such waves, each with a different wavelength, is also a solution. So consider a bunch of different waves with different wavelengths. Each of the waves begins somewhere west of Los Angeles and proceeds through Kansas City to somewhere east of Hoboken, New Jersey, oscillating to and fro in a manner given by its wavelength. But envision the following: Each wave is arranged such that it has a peak at exactly the same place in a steakhouse in downtown Kansas City. All of the many waves are small sounds, but they all happen to add up at exactly that point in Kansas City. This would produce an enormous "boom" that would blow the roof off the steakhouse (if the hot sauce on the rib eye hasn't done that already).

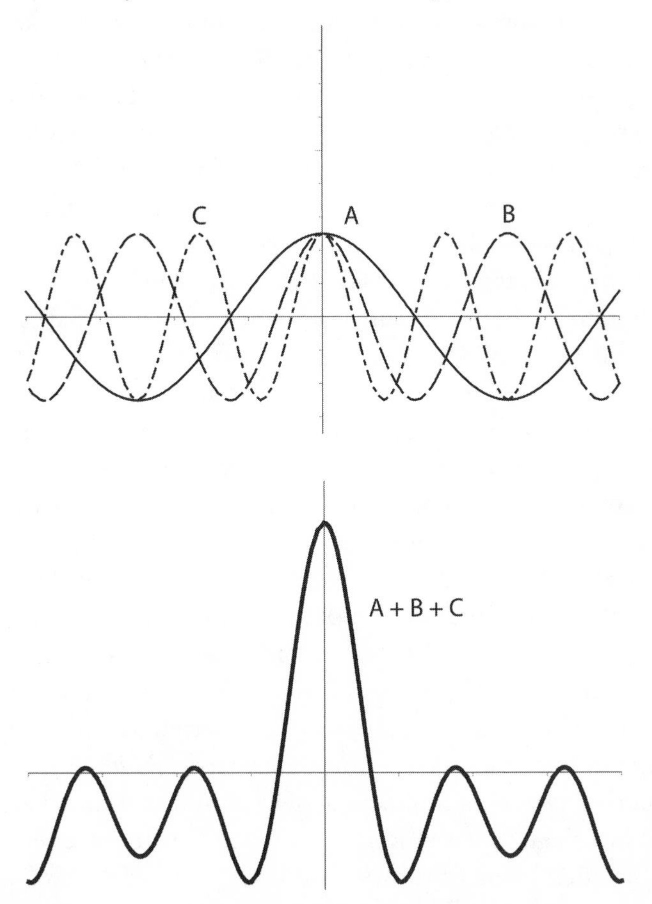

FIGURE 22: By combining normal traveling waves together, for example, (A = cos(x)), (B = cos (2x)), and (C = cos (3x)), we make a wave (A + B + C) with a localized bump. The bump occurs where all three constituent waves are exactly in phase, in this case at the origin. It trails off where the constituent waves are out of phase. Schrödinger thought this gave rise to the localized "particle" concept, but in fact the particle is everywhere with a larger probability of being found where the (square of the) wave is greatest.

What Fourier analysis shows us is that if we combine enough waves together, of the appropriate amount for each wavelength, then there can be an enormous lump for the total sum, located somewhere in space, such as in downtown Kansas City. But it can be arranged so that, as we go east and west from there, the crests and troughs of the waves would all coincide and would cancel out completely. So this series of waves, each one of which goes on forever, when added together, has a significant value only at one small place in space, our favorite restaurant in Kansas City, and is zero everywhere else (we've effectively engineered the peak of an enormous "rogue wave" in downtown Kansas City). And, if all the waves we added together are moving, like water waves, from west to east, then the location of the place where they all add up to a lump also moves from west to east. This mathematics of adding all the moving waves, however, has one curious consequence: each of the constituent waves, with different wavelengths, moves with a slightly different velocity. Therefore, the careful arrangement of all these waves to make one well-localized lump at one place in space soon begins to deteriorate. The narrow pulse describing our well-localized particle begins to broaden in time, and consequently our knowledge of where the particle is located slowly deteriorates. The particle lump begins to fade away.

All this applies to the Schrödinger wave function. When we localize a particle at a point in space, for example, $x = 0$, at a time $t = 0$, it is like throwing a rock into the lake. We get a big splash where the rock disturbs the water at that particular point in space and time. We have a Fourier sum of many waves adding up and producing a big bump in $\Psi(x, t)$ at that special point ($x = 0$ and $t = 0$). But the splash begins to disperse and, over time, just appears to be a distribution of many waves propagating out into space. Where has the "particle" gone?

The Schrödinger equation was a success almost immediately as it reproduced Bohr's and Heisenberg's results for the atomic energy levels, but it had the added benefit of bringing back a true picture of what the atomic energy states actually looked like. What Bohr thought of as circular orbits were now fuzzy little "orbitals," $\Psi(x, t)$. Here the electrons were bound into the atom, and $\Psi(x, t)$ didn't disperse out into space. These bound electrons had wave functions that were like the oscillating modes on a stringed instrument. It should be noted that the motion of *any particle* that is localized, or trapped by a binding force, will behave like the waves on a stringed instrument and have corresponding energy levels that are *quantized* like the plucked notes of the string and will take on only particular discrete allowed values.[17] This happens to electrons bound in atoms, as well as to protons and neutrons bound in the atomic nucleus, and to quarks, which are bound inside of protons and neutrons. In the case of quarks bound within particles, the energy levels that represent the excited states of motion of the quarks actually appear to us as new massive particles. And, finally, *string theory* is a relativistically glamorized version of a guitar string. The goal of string theory is to explain the quarks themselves (and all the other truly fundamental particles in nature) as the quantum vibrations of a string. Such wonderful music can be heard from that old guitar, if one simply practices.

Schrödinger's construction was called "wave mechanics," a theory that J. Robert Oppenheimer, the director of the Manhattan Project, called "one of the most perfect, most accurate, and most lovely theories man had discovered."[18] Unlike Heisenberg's matrices, it dealt with a more familiar mathematics (familiar to most physicists of the day as differential equations). Schrödinger thought he had brought sanity to the understanding of the atom and the nascent quantum theory. It now resembled classical physics. There were no particles, only waves that, by superposition, could look like localized particles.

But, alas, this wasn't quite the right way to think about the quantum world. Quantum mechanics is the proverbial mysterious entity that is being felt by the blind. While Heisenberg had described the tusks and Schrödinger the trunk, the total elephant is so much more than its parts.

WAVES OF PROBABILITY

Here's part of the problem. Let's say a wave function, $\Psi(x, t)$, represents a set of waves that add up to a lump—an electron located within a small space and moving with a particular velocity. When this wave function—a Fourier sum of waves—collides with a barrier, part of it will be reflected and part of it transmitted. The math is clear on this point. The wave function that initially is a lump divides into two lumps, one reflected from the barrier and one transmitted through the barrier. But an electron is not breakable into two parts!

Either the electron is reflected or it is transmitted through the barrier: this is a fact, and it can be tested experimentally. You'll never find 10 percent of an electron going through the barrier and 90 percent of an electron reflected.

A contemporary colleague of Schrödinger, Max Born, who worked in the 1920s with Wolfgang Pauli and Werner Heisenberg at the University of Göttingen, realized that the naive idea of "matter waves" that approximate the shape of a particle doesn't work as an interpretation of Schrödinger's wave function.[19] Particles are "digital," either one whole particle is detected or none at all; waves are fuzzy and wavy, which led some naively to the idea of measuring fractional parts of particles—something that doesn't match reality. Born provided a physical interpretation of the wave function that has both empowered and haunted quantum mechanics ever since. Strongly influenced by Heisenberg's uncertainty principle, Born proposed that the (absolute) square of the wave function, which is always a real number and is always a positive number, is *the probability of finding the particle* at any given point in space, at any particular time:[20]

$$\Psi(x, t)^2 = \text{probability of finding particle at position } x \text{ at the time } t$$

Born's interpretation of Schrödinger's wave function thus locks together, inextricably, the notion of a particle to the notion of wave. It is also terrifying, or humiliating, depending on one's perspective—physics now had to deal with *probability as a fundamental component of a physical theory*. We simply can no longer make exact statements about the familiar positions and motions of things. We must be content, according to the very laws of physics, with more limited information about the outcome of a physics experiment.

Unlike the language of Newton or Einstein, we cannot talk about the exact position of the particle $x(t)$ at the time t. Rather, all our available information is now encoded into $\Psi(x, t)$, the value of the quantum wave function at position x at the time t, and only its absolute square is measurable.

Incidentally, it was Max Born who coined the term "quantum mechanics."[21] Born had put his finger on what the Schrödinger equation and $\Psi(x, t)$ is really describing—for there was no doubt there was some sort of wavelike behavior associated with particles—it had, after all, been seen at Bell Labs for electrons. The wave function $\Psi(x, t)$ represented (the square root of) a wave of probability. Where $\Psi(x, t)^2$ is large, there is a large probability of finding an electron; where $\Psi(x, t)^2 = 0$, the electron never turns up. The wave function $\Psi(x, t)$ can oscillate, taking on any (complex)[22] value at any point in space and time, but "probability" can only have positive values between zero and one. So Born interpreted $\Psi(x, t)^2$ to be the probability, and he added the warning label on the Schrödinger equation that the total probability, for example, of finding the electron to be somewhere in space at any given time must always be equal to one.[23] The probability distribution, $\Psi(x, t)^2$, could have wavelike properties, but the electron itself is a real, solid particle. The interpretation of the barrier problem (the Victoria's Secret window) then becomes one of statistics. If the Schrödinger equation predicts that 90 percent of the wave is reflected and 10 percent transmitted, this means that out of a thousand electrons, something like nine hundred will be reflected and one hundred transmitted. But what happens to a single electron? To find out its fate, we must roll the dice—in this case, a ten-faced die with nine faces having "Rs" (meaning "reflected") and one face having "T" (transmitted). At least, this is what nature seems to do: *nature does roll dice*, permitting humans to predict only probabilities for the results of experiments at the quantum level.

Born's interpretation of the Schrödinger wave function was actually inspired by a 1911 paper of Einstein, but in 1926 it represented a scientific and philosophical upheaval that was nothing less than an intellectual Armegeddon. After the old Newtonian world of absolute certainties, it was not easy to accept a Mother Nature who yields mere probabilities for whatever things you want to measure, or predict, whether it be a particle's location, velocity, energy, and so on. Schrödinger, for one, rejected it vehemently, regretting having devised the equation that gave rise to it.

Sigh. Now surely everything has been cleared up. Or has it? As you may have guessed, clarity is not so easy to come by when it comes to quantum physics. Remember that, at this point, observation after observation has presented us with absurd contradictions. A major resolution did arrive in the years 1925–1927, thanks to a series of stunning intellectual breakthroughs by our gang of intrepid quantum explorers. This included the aforementioned Erwin Schrödinger, Werner Heisenberg,

Max Born, and our deep-thinking Dane, Niels Bohr, as well as the painfully shy Paul Dirac, the irascible critic Wolfgang Pauli, the learned mathematician Pascual Jordan, and let's not forget the contributions of Einstein, Planck, and de Broglie—a stellar group, if there ever was one.

THE TRIUMPH OF UNCERTAINTY

It is, in all its splendor, the "uncertainty relation" that made Heisenberg a household name:

$$\Delta x \, \Delta p > \hbar/2$$

It spells out, among other things, why an electron's precise path between two point, A and B, forever eludes us. Let's take Δx (call it "delta x"), which is a symbolic way of representing the best we can possibly do, the closest we can come, or the residual uncertainty, in measuring the position of an electron along the x-axis of coordinates.[24] Δp is the corresponding uncertainty in the other quantity we need to know when we are trying to pin down a particle, namely, its momentum, p, also along the x-axis.

What Heisenberg discovered was that any act of measurement that yields a small Δx (which brings us pretty close to the particle's location) will, unfortunately, generate a correspondingly large Δp, a huge uncertainty in momentum. If we manage to shrink Δx to zero (no uncertainty in location), our Δp (equivalently, its velocity) swells to infinity, for Δp must be infinitely large for the product to exceed Planck's constant (the h in the formula). Thus, these two quantities—position uncertainty along some axis and the corresponding momentum uncertainty along the same axis—are forever locked in a reciprocal embrace.

It might have been more apt to call Heisenberg's discovery the "unknowability relation," since it tells us that there are certain things that are intrinsically unknowable. The whole equation says that the smallest interval in our knowledge of position, multiplied by the smallest interval in our knowledge of momentum, of an electron, must be greater than or equal to Planck's constant divided by 2. (Note the reappearance of the quantum domain logo h, which we first encountered in the famous formula of Planck, $E = hf$. Incidentally, the quantity $\hbar = h/2\pi$ is pronounced "aitch bar," and it appears so frequently in equations that it was give its own symbol. Often physicists will say "Planck's constant" when they mean $\hbar$.) The gist of it, again, is that the more we know about the position of a particle, the less we know about its momentum, and vice versa. What Heisenberg tells us is that no matter how good our equipment is, Mother Nature has so arranged the microuniverse that the product of the two uncertainties always exceeds Planck's constant. (In the classical realm we can discount this pesky h, which is dwarfed by objects billions of times larger, so we can be exact in our measurements of the trajectories of baseballs, planets, and Porsches.)

Time, Einstein showed in 1905, stands in relation to the three space coordinates, x, y, z, in a four-dimensional "space-time." The other ingredient in nature's fuzziness, Heisenberg found, can be expressed as $\Delta E \Delta t > \hbar/2$. This means that in the quantum world, the two properties, energy and time, also refuse to be pinned down simultaneously. The more precisely you know when a particle passed through, for example, a given slit in time, the less you precisely you can know its energy, and vice versa.

BORN, FOURIER, AND SCHRÖDINGER

Max Born's probabilistic interpretation of $\Psi(x, t)$ has built into it the essence of Heisenberg's uncertainty relations. Let's take the simplest form of the Schrödinger equation, which describes a single particle, say, an electron, moving with some specific velocity. The solution is a single wave stretching over all space with a defined wavelength (which, you may recall, was set by Louis de Broglie to be equal to the momentum divided by Planck's constant). So we know everything about this electron's wavelength (or its momentum, since the two are related) but we know nothing about where it is. If the electron is traveling along the x-axis, its location can be anywhere on that axis, from negative infinity to positive infinity. That's quantum science. If we know the motion (momentum) precisely, we know nothing about the location.

But what does the Schrödinger equation tell us about an electron whose location is more precisely known? Here is the beauty of mathematician Fourier's idea. Remember our particle in Kansas City? A localized disturbance can be represented mathematically by a sum of infinitely extended waves, each with a different wavelength. This is pure math, so the disturbance can be anything: a pulse of sound (acoustic waves), a moving bump on a long rope, an electrical pulse of voltage on a long wire, a rogue wave crest at sea—all these and more make good use of Fourier's analysis. In each case, the detailed shape of the localized disturbance that we want to describe determines the number of waves and the range of wavelengths that must be added up.

Fourier showed that the more localized the pulse, the larger the band of wavelengths that are required. A modern example is a high-fidelity acoustic system, which, in order to faithfully transmit very short pulses of sound, must have a wide frequency acceptance since a large band of wavelengths is needed to describe a short-duration sound pulse. So what has this to do with Schrödinger's equation? Remember that Schrödinger's wave function waves were interpreted by Max Born as probability waves. And if we know the electron's location, we can think of it as a "pulse of probability," as in Fourier's math.

Now comes the point. If the electron's location is known, then the uncertainty is small. Using Fourier, we can see that the band of wavelengths of waves needed to describe this narrow pulse of probability is correspondingly large. In other words, Fourier's two hundred-year-old equations back up Heisenberg's uncertainty relationship. If we know "pretty much" where the electron is, Heisenberg says we know next to nothing about its momentum. Likewise, the electron's location can be described as a sharp "pulse of probability," in which case Fourier says we need a large number of wavelengths, which in turn implies a large uncertainty in the knowledge of the momentum. Over a century earlier, Fourier provided the mathematics for just what Heisenberg was describing.

THE COPENHAGEN INTERPRETATION

In early 2000, a play opened on Broadway, imported from England, called *Copenhagen*, by Michael Frayn. It has just three characters, Niels Bohr; Bohr's wife, Margrethe; and Werner Heisenberg. The play portrays a historically correct visit of Heisenberg to Bohr's lab in Copenhagen during World War II, when Germany occupied Denmark. Heisenberg, then a leading German scientist, was involved in war work in Nazi Germany. The substance of the meeting is not known, but the playwright portrayed it as a fascinating mixture of politics and science.

Heisenberg's actual role in the attempt to build an atomic bomb is not known with certainty. Some historians have speculated that he deliberately forestalled any real effort to succeed in building a nuclear bomb, while others claim he failed to grasp the technical aspects of such a weapon. Curiously, Heisenberg's role was as enigmatic and uncertain as his famous uncertainty relation, which became the foundation of the new quantum science.

But why these particular pairs of opposites? Why is it that the more we know about a particle's location, the less we know about where it is going (momentum), and ditto about energy and time? Bohr called these quantities complementary variables because knowledge of one limits knowledge of the other. It is as if we were trying to learn about an Oriental rug by studying the details of the weave and the overall pattern: To analyze the weave, we would focus closely on it, thereby losing track of the overall pattern. To see the overall pattern, we would have to step back a bit, losing the details of the weave.

We have mentioned that Heisenberg, like Bohr, banned all statements that were not capable of being experimentally verified. Thus, when it came to Δx, he imagined a variety of devices to measure the position, x, for an electron. These were "*gedanken* experiments," or "thought experiments"—imaginary but plausible physical experiments that were and are used by theoretical physicists. Thought experiments don't involve getting one's hands dirty, but they may involve staying up all night to try to figure out theoretically what the outcome is.

One *gedanken* experiment was a "gamma ray microscope" based on accepted principles of optics. Because Heisenberg wanted good precision (a small Δx), he insisted on short-wavelength light—therefore gamma rays, the electromagnetic waves of the shortest wavelengths. In so sharpening his knowledge of the coordinate x, however, the energetic gamma rays would have greatly disrupted the electron, changing its momentum, p, by a large and unknowable amount. Example after example showed a consistency with the uncertainty relations, where the quantum logo, h, entered because of the de Broglie relation between wavelength and momentum. The smallest possible uncertainty of position and the smallest possible uncertainty of momentum must generate a product somewhat greater than $\hbar$.

So we must swallow hard and accept the fact that reality in the quantum domain is probabilistic. There is also probability in classical physics, in dealing with large numbers of particles whose positions and momenta are too numerous to record. But here the uncertainties could be reduced in microscopic experiments to negligible values so that our predictions of future outcomes would be virtual certainties. Thus we can predict that Jupiter is not going to careen into Saturn next week. But in quantum physics the uncertainties are always there and are "hardwired" into the laws of nature.

Bohr took this a step further, in what became known as the "Copenhagen intepretation" of quantum mechanics. It is meaningless, he said, to try to picture the trajectory of an electron. What can't be measured does not exist. Despite the reassuring cloud chamber portraits of electron trajectories, the reality is that the very concept of particles traveling defined paths is misleading and erroneous. Bohr's magical circular atomic orbits do not exist, after all. He finally asserted that only probabilities can be known.

This is shocking. Human brains are not naturally designed for quantum reality, and so it is natural to look for a way out of this queasy uncertainty. Over the years, many great physicists have tried to fight the implications of Heisenberg. Einstein, who loathed the probabilistic nature of reality, devised many ingenious thought experiments to circumvent this "inevitable disturbance." And his sparring with Bohr was a delightful chapter in the history of quantum theory—well, delightful for Bohr, maybe not so for Einstein. As we shall see, all the master's schemes ultimately ran aground against the irreducible conclusion that indeterminacy is intrinsic to the atomic domain.

So what does the Copenhagen interpretation tell us about the enigma of the double-slit experiment enigma; in other words, which path does the electron take? There is no problem: the probability waves interfere, and electrons appear where the probability is large.

STILL CRAZY AFTER ALL THESE YEARS

So where are we, after Heisenberg, Schrödinger, Bohr, Born, and all the others? Now we have probability waves and uncertainty relations as a way of maintaining the particle viewpoint. The crisis of "sometimes a wave, sometimes a particle" is resolved. Electrons and photons are particles. The description of their behavior involves probability waves. These can interfere, and the obedient particles appear where they must, in accordance with the probability wave function. How they get there is not a permitted question. That's Copenhagen. The price of this success is probability and quantum weirdness.

The notion that nature (or God) plays dice with subatomic stuff never sat well with Einstein, Schrödinger, de Broglie, Planck, and others. Einstein cherished the belief that quantum theory was merely a stopgap, which would eventually be replaced by a theory that was deterministic and causal. Over the years, he made many clever attempts to show that uncertainty relations could be circumvented, but they were foiled, one by one, with relish, by Bohr.

So we close this chapter with a mixture of triumph and lingering existential unease. By the end of the 1920s, quantum mechanics had come of age, but new successes and further refinements would keep it cooking well into the 1940s.

Chapter 6

QUANTUM SCIENCE AT WORK

Otherworldly, though it may seem, the quantum theory of Heisenberg and Schrödinger actually worked miracles! The picture of the hydrogen atom now came into focus without the mental crutch of Keplerian planetary orbits but with the novel fuzzy Schrödinger wave functions, now called the electron "orbitals." The new quantum mechanics became a tool of tremendous power as physicists became increasingly adept at applying Schrödinger's equation to various domains and ever-more-complex atomic and subatomic systems. As Heinz Pagels wrote, "The theory released the intellectual energies of thousands of young scientists in the industrialized nations of the world. No single set of ideas has ever had a greater impact on technology and its practical implications will continue to shape the social and political destiny of our civilization."[1]

Still, when we say "it works," about a scientific theory or model, what exactly do we mean? We mean that the theory, through the use of mathematics, makes certain assertions about nature and that these assertions can be compared to our accumulated experience. If these assertions and the experiences agree, the theory works in what we call the "postdicting" mode: explaining what we already know to be true but didn't understand before.

For example, we might drop two objects of different mass off the Leaning Tower of Pisa. Galileo's demonstration and all our subsequent experiments showed that, apart from the small corrections due to air resistance, both objects, if dropped from the same height, will fall to the ground in the same amount of time. This is exactly true when there is no air resistance, as on the surface of the moon. This was dramatically demonstrated on TV by a lunar astronaut dropping a feather and a hammer simultaneously, both hitting the lunar soil at the same instant.[2] The new and deeper theory to be tested in this case is Newton's law of motion, that is, that the force on an object equals its mass times its acceleration. Newton's fame also rests on his universal law of gravitation. When these two principles are combined, we can predict the motion of the falling objects, and predict how long it will take for two objects dropped from the same height to reach the ground. Newton's theory neatly explains the fact that objects arrive at the ground simultaneously (if we neglect the effects of air resistance).[3]

But a good theory must also be able to *predict* what will happen if we do something that has never been done before. When the ECHO satellite was launched into space in 1958, Newton's theory was used to predict its path, given the forces of the rocket motor, the force of gravity, and corrections for other important factors, such as wind velocity, Earth's rotation, and so on. The predictive power of equations depends, of course, on how much control we have over all the determining factors. Again, we have witnessed spectacular success of the theory. Newton's theory correctly postdicted the world, and it equally well predicted it over the vast domain of velocities (less than the speed of light) and distance scales (things bigger than atoms) to which it applies.

BUT ISAAC NEWTON NEVER SENT US E-MAIL!

Now let us ask: Does the quantum theory explain ("postdict") the world we live in? Can it be used to predict new phenomena never seen before and thereby be instrumental in inventing new and useful devices? The answer to these questions is a resounding yes! In countless tests, both pre- and postdictive, the quantum theory has proved a resounding success. It departs from its predecessor's theories, Newton's mechanics, and Maxwell's electromagnetism, whenever the "quantum logo"—Planck's famous and very small constant h(or $\hbar$)—cannot be ignored in the equations. That is the case when the masses, sizes, and time scales of the objects being described are comparable to those of atoms. And since everything is made of atoms, we should not be surprised that atomic phenomena can occasionally rear their heads in the macroscopic world of people and their measuring instruments.

In this chapter we will explore this spooky theory, and how really eerie it is will soon become even clearer. It will explain all of chemistry, from the periodic table of the elements to the forces between atoms that create molecules (which the chemists call "chemical compounds," and of which there are billions). Then we will explore how quantum physics materially affects almost every corner of our lives. God may play dice with the universe, but man has managed to control the quantum domain enough to fashion the transistor, tunnel diodes, die lasers, x-ray machines, synchrotron light sources, radioactive tracers, scanning tunneling microscopes, superconducting magnets, positron emission tomography, superfluid liquids, nuclear reactors and nuclear bombs, MRI machines, the microchip, and the laser—to name just a few. Although you may not have any superconducting magnets or scanning tunneling microscopes around the house—you probably do have a hundred million transistors. Moreover, your life is touched every day by all the things that quantum physics has made possible. If we had been stuck in a purely Newtonian universe, there would be no Internet to surf, no software wars, and no Steve Jobs or Bill Gates (or rather, they might have been railroad tycoons). We may not have had some of the modern problems we now face, but we certainly would not have had the tools to solve the many problems humanity faces today.

The implications for all of science, beyond the walls of the physics department, are equally profound. Erwin Schrödinger, who gave us that exquisite equation governing the entire quantum realm, wrote a prescient book in 1944 called *What Is Life? Mind and Matter*.[4] The book made a guess as to how genetic information works. When the young James Watson read this remarkable book, it piqued his interest in DNA, and the rest is history: Watson, with Francis Crick, went on to unravel the double helix structure of the DNA molecule, thus launching the revolution in molecular biology of the 1950s and, in time, the bold new era of genetic engineering in which we now live. Without the quantum revolution we would not have been able to understand the structure of any molecule, let alone the DNA molecule, the basis of molecular biology—indeed, the very basis of life.[5] On the more far-out and speculative frontier, explaining the problem of the mind, human self-awareness, and consciousness may require the richness of quantum science, according to some fearless theoretical physicists who dare to tackle cognitive science.[6]

Quantum mechanics continues to illuminate chemical processes: The 1998 Nobel Prize in Chemistry, for example, was awarded to two physicists, Walter Kohn and John Pople, who developed powerful computational techniques to solve the quantum mechanical equations that determine the shapes and interactions of molecules. Astrophysics, nuclear science, cryptography, materials science, electronics, as well as chemistry, biology, biochemistry, and so on, would be likewise impoverished without the quantum revolution. Information technology would exist only as the science of designing file cabinets for paper documents without quantum physics. Where would this field be without Werner Heisenberg's uncertainties and Max Born's probabilities?

Without quantum theory, the pattern and the properties of the chemical elements—as embodied in the periodic table of elements that predated quantum theory by half a century and which defines all chemical reactions and chemical structures and gives rise to all things in our lives and life itself—would never have been fully understood.

PLAYING CARDS WITH DMITRY MENDELEYEV

However, chemistry, like physics, was already a respectable and evolving science long before quantum theory arrived on the scene. In fact, it was through chemistry that the reality of atoms was established by John Dalton in 1803, and it was through Michael Faraday's research in electrochemistry that the essential electrical nature of atoms was demonstrated. But atoms were not understood at all. Quantum physics would now provide chemists with a profound and rational explanation for the detailed structure and behavior of atoms as well as a formalism for understanding and actually predicting the formation and properties of molecules. It is the very probabilistic nature of quantum theory that yields these successes.

Yes, indeed, chemistry was not everyone's favorite subject, even if it does dominate much of modern technology. Willie may have written, on his high school chemistry final, that H_2O is hot water and CO_2 is cold water. It is our belief that if you explore with us some of the logic behind chemistry, you will be hooked. You will find that the unraveling of the atom is one of the greatest forensic detective stories in human history.

Chemistry surely begins with that famous chart, the periodic table of the elements, which graces the walls of hundreds of thousands of chemistry classrooms the world over. The periodic table was a truly major development, based on a law of chemistry discovered by the startlingly prolific Russian chemist Dmitry Ivanovich Mendeleyev (1834–1907). Mendeleyev thrived in czarist Russia. He was a prodigious scholar, authoring over four hundred books and articles, and a "practical" chemist, contributing to such diverse fields as fertilizers, cooperative cheese making, weights and measures, Russian trade and tariffs, and shipbuilding. In the meantime he supported radical student causes, divorced his wife and married a young art student, and, as his picture would indicate, permitted himself only one haircut a year.[7]

Mendeleyev's chart is based on writing down a sequence of the atoms by their increasing atomic weight. Please note that by "element" we mean a particular atom or a material containing one particular kind of atom. Thus, a block of the element carbon would be a block of graphite or a diamond; both contain exclusively atoms of carbon, although the atoms are arranged in different ways, causing graphite to be dark and useful in pencils, while diamond is hard and impressive to would-be fiancées, not to mention useful in drilling through hard metals. In contrast, water is not an element; rather, it is *composed* of the elements hydrogen and oxygen, which are fastened together by electrical forces. They are also governed by Schrödinger's equation, so we say that water is a *compound*.

The "atomic weight" of an atom is simply its mass. Every atom has a distinct mass. All oxygen atoms have the same mass. All nitrogen atoms also have the same mass, but it is different than oxygen (nitrogen is a little lighter than oxygen), and so forth. Some atoms are very light (low mass)—hydrogen being the lightest—while others are hundreds of times heavier, like uranium. The mass of an atom is most conveniently measured in certain special units,[8] but their precise values and determination aren't important for our purposes now. Still, we are interested in making a list that is a sequence of the increasing atomic masses of the atoms. Mendeleyev observed that the position of an atom in this sequence has a clear relationship to the element's chemical properties. This was the key to unlocking chemistry.

POLICE LINEUP OF THE ELEMENTS

So let's think of a police lineup in which the suspects are arranged from left to right, starting on the left with the puniest lightest-weight gangster. Each suspect has a name, and we abbreviate the names: so hydrogen is just "H," while oxygen is "O," and iron is "Fe" (coming from "fer," which is Latin for *iron*), helium is "He," and so forth. However, the suspects are lined up, not alphabetically, but in increasing weight as you go from left to right.

We start with the lightweights on the left, then the welterweights come next, and then we continue on up to the biggest heavyweights way down the line to the right. This is an "ordered sequence" of increasing atomic masses. The position of a particular atom in this sequence is called the "atomic number." We call the atomic number "Z."

So, we see that the puniest suspect, the hydrogen atom, which is the lightest atom (smallest atomic weight, approximately A = 1 in certain units),[9] is therefore assigned the atomic number Z = 1, since it is first in the lineup. The helium atom weighs in second lightest (with approximate atomic weight, A = 4, almost four times heavier than hydrogen), and because it is the second suspect in the lineup, we say that helium has Z = 2. Next comes lithium (with weight A = 7), and so it has Z = 3; then beryllium, Z = 4; and so forth. The police lineup of the lightest several atoms is shown in figure 23, but the lineup today goes well beyond one hundred atoms.

The atomic number, Z, is the main identifier of any atom. Z measures where in the lineup a given suspect is to be found. If you say, "who is suspect number Z = 13?" the answer is "aluminum" (nicknamed "Al"). And who is suspect number Z = 26? "Iron" (nicknamed "Fer"). You should spend a moment or two perusing the suspects and who has what atomic number. Remember, Z is not the atomic weight but rather just the numerical order in the sequence of lightweight to heavyweight.

Z is of supreme importance in atomic physics. Skipping ahead for a moment, quantum theory will teach us that Z is *actually the number of electrons orbiting that atom's nucleus*. So, for example, sodium (Na) has atomic number 11; therefore, 11 electrons orbit the nucleus of a sodium atom. The atoms are all electrically neutral, thus there must be an equal but opposite (positive) electric charge attached to the nucleus and that

FIGURE 23: The witness, Mrs. Fenster, accompanied by Officer O'Reardon, inspects the lineup of suspects (which continues up to 118 in total). The suspects are ordered in increasing weight to the right. Notice, however, that both Li and Na are wearing the same star-spangled Elvis jackets, a clear hint of a conspiracy. (Illustration by Ilse Lund.)

must also equal the atomic number Z. Hence, sodium has 11 electrons orbiting around in fuzzy quantum style and 11 positive charges deep in its interior, lurking within the dense and compact nucleus (recall the discovery of the nucleus by Ernest Rutherford). Today we know that the positive charges are protons in the nucleus of the atom. So sodium is Z = 11, eleventh in the lineup, with 11 electrons orbiting the nucleus and 11 protons deep inside, balancing the electric charge. The sodium atom, like all atoms, has a net electric charge of zero. So much information is contained in Z. But it is perhaps hard to appreciate that *none of this* detailed internal atomic structure was known to Mendeleyev. He didn't have knowledge of these inner workings, but like a great detective he simply began with his police lineup.

Mendeleyev then discovered a remarkable pattern in the chemical behavior of the elements as they participated in various chemical reactions and as their atomic number, Z, increased. The chemical behavior of elements is *periodic*. That is, as we start with a given atom and go upward in Z, we eventually get back to a much heavier atom that behaves almost the same way as the atom we started with! There is a hidden inner conspiracy among the suspects in our police lineup: different suspects behave almost exactly the same way. The precise pattern was refined as more atoms were discovered and added to the lineup. Indeed, the periodic behavior of atoms actually led to the discovery of many new "missing" elements that were needed to complete the sequence. Let's see what the conspiracy is in the modern sequence of atoms. (Though many of the atoms were unknown to Dmitry, we will proceed in the Mendeleyevian manner.)[10]

First let us think about these "chemical behaviors." What are chemical behaviors or chemical properties? We all know that we can easily dissolve things such as salt in water, but oil does not dissolve in water. Also, water doesn't burn (it puts out most fires), while things like carbon (Z = 6), which is the main ingredient in coal, easily burn, and things like iron (Z = 26) rust, in a slow kind of burning. Take away oxygen (Z = 8), then none of these things burn or rust. In fact, burning or rusting is just *oxidation*, the chemical combination of oxygen with another element. Some elements we see will readily combine with oxygen, while others will not. This has to do with energy (when two oxygen atoms combine with a carbon atom we get CO_2 [carbon dioxide] and a certain amount of energy is released), but it also has to do with the specific chemical properties of carbon and oxygen atoms that allow this to happen. We breathe oxygen, and it keeps us alive by oxidation reactions thoughout all the cells in our body (we are sort of burning!), but we get no breathing benefit from nitrogen (Z = 7), even though oxygen and nitrogen stand right next to each other in the police lineup. These are all familiar and basic chemical behaviors. But what makes one atom, oxygen, such a great oxidizer, and nitrogen not?

For a little more detail about chemical properties, let's consider the element lithium, with Z = 3. Pure lithium (Li) is a shiny, malleable metal, but it rapidly reacts when exposed to moisture in the air, forming a layer of lithium hydroxide, LiOH. Usually the metal is stored in oil to block the humidity in air from contaminating the surface (incidentally, lithium has such a small atomic weight that it has a small density and it actually floats on water). Now, if a chunk of the metal lithium is thrown into water, H_2O, a spectacular and violent set of chemical reactions occur, releasing hydrogen and a lot of energy. This results in the hydrogen burning (or exploding) in the air above the water, as seen in two of many links to various YouTube videos of lithium-water and sodium-water fires.[11] In summary, water rapidly reacts with lithium, forming lithium hydroxide, LiOH, and releases hydrogen gas into the air. Because of the explosive fire that usually occurs at the water's surface, where hydrogen and atmospheric oxygen, O_2, are rapidly ignited, we don't dispose of metallic lithium by throwing it into water.

So here is the conspiracy that Mendeleyev noted about the elements in the lineup: If we now go up by eight steps in our police lineup, the Z sequence, from lithium with Z = 3, we arrive at sodium (Na, which in German is called *Natrium*) with Z = 3 + 8 = 11. Sodium is also a shiny metal that rapidly turns a whitish-grey color on its surface when exposed to the air (forming a crust of sodium hydroxide, NaOH, on its surface by reacting with humidity in the air). If you throw sodium in water, guess what happens? A rapid and violent reaction occurs with the liberation of hydrogen gas that usually ignites in the air, and you get an impressive explosion. Sound familiar? Sodium is a much heavier atom than lithium, but chemically it behaves the same way. How could that be? Lithium and sodium are eight steps apart on the police lineup, yet they seem to be in the same gang. This is one instance where it is advisable to invoke a conspiracy theory. There is an evident conspiracy of similar behaviors among the suspects in our police lineup.

In considering the vast number of chemical reactions in nature, we find that Li and Na are quite similar. They react slightly differently, for example, with slightly different reaction rates, as expected because the Na atom is heavier than the Li atom, but almost any chemical compound (molecule) you can make containing Li can be modified by substituting the Na atom in its place, and vice versa.

Now, if we go up another eight steps in atomic mass we arrive at potassium (K). And you likely guessed it. Potassium behaves the same way as sodium and lithium. In fact, though it is a gas at normal temperatures and pressures, even hydrogen behaves much like these elements, permitting chemical substitution. So H_2 gas can easily be chemically rearranged, by replacing an H with an Li, to become LiH (lithium hydride), NaH (sodium hydride), KH (potassium hydride) and so on, or H_2O = HOH can be rearranged to become LiOH (lithum hydroxide), NaOH, (sodium hydroxide), KOH (potassium hydroxide), and so on. Clearly, hydrogen, lithium, sodium, and potassium are all in the same gang!

While we have now observed (as, originally, did Mendeleyev) from the police record the remarkable similarity of certain of the atomic suspects, we still don't understand why they're so similar. Why the magic number 8? Nor do we understand what controls these remarkable and often violent chemical reactions.

When scientists are confused, they first classify things, then give them fancy names. So we give the gangs in the lineup special names: we call the gang that contains H, Li, Na, and K the "alkali metals." (Yes, we know, hydrogen isn't a metal, but that's the name of the gang that the puniest guy is in.) In Mendeleyev's style, much like the local police authorities might do with gangsters, we make a vertical list and we put the names of the alkali elements in it (figure 24).

This forms the first column of the periodic table, where the members of a single gang of atoms with similar chemical properties are listed now in ascending order in

1 H
3 Li
11 Na
19 K
37 Rb
55 Cs
87 Fr

FIGURE 24: The "alkali metals" are a chemical family of atoms or elements, each having the same chemical properties.

their weights. As noted, this is a list of the particular gang in the police lineup called the "alkali metals."

If we ascend further in Z beyond potassium, the pattern astonishingly changes. To get to the next alkali metal (same chemical behavior as the others), we must ascend to rubidium (Rb) 18 = 8 + 10 steps above potassium. (Rubidium confessed to being a member of the alkali gang only after much chemical interrogation.) Then another eighteen steps gets to cesium (Cs), then 32 = 8 + 10 + 14 steps to get to the last alkali metal, francium (Fr). All these heavyweights also belong in the "alkali metals" gang. They comprise the entire first column of the periodic table, which ends with francium. But why did the magic number 8 change to 18? Then to 32? What is going on here? And why does the list end with francium?

The last question is related to the overall stability of atoms, mainly the stability of very heavy atomic nuclei. It turns out that as atoms get heavier, like francium, their nuclei become very unstable and thus radioactive. For the heaviest atoms, the nuclei have become so unstable that they cannot exist for longer than very brief fractions of a second in the lab (this is not chemistry anymore—it's now nuclear physics). Francium is only produced on Earth by the radioactive disintegration of heavier atoms, such as uranium, and then it quickly disintegrates by its own radioactivity. At any instant it is estimated that there is only about one ounce of francium on the whole of planet Earth, and it is the rarest naturally occurring element of all. But that's enough to determine its chemistry, and so it is found to be an alkali metal. We have now constructed the first column of the periodic table of the elements (figure 24) and we have witnessed the conspiracy, or pattern of chemical periodicity, first seen by Mendeleyev. Such a weird pattern! What is going on? Have no fear—quantum theory will explain it all.

In fact, quantum physics explains why all the chemical elements belong to various gangs that form the various columns in the periodic table of the elements. Consider helium (He). Helium was not known in Mendeleyev's time because it doesn't react chemically with anything. It is a very light atom, Z = 2, that occurs only as a gas. Being lighter than O_2 and N_2, it rises in the atmosphere and eventually escapes from the planet into outer space. Helium is great for dirigibles and blimps because it won't burn. The sun (and other stars) is composed mostly of hydrogen and helium. Helium's lightness in weight makes our voices sound like Donald Duck if we inhale it and then speak. Since helium can't react chemically with anything, this is a fairly harmless trick for a reasonably healthy adult to perform (but we don't recommend doing this with hydrogen, at least not while smoking a lit cigarette, because it would explode, perhaps in your lungs). Since helium won't react chemically, that is, it won't form compounds with anything, we say that helium is "chemically inert."

Again, if we go up eight steps in atomic mass from He, we arrive at neon (Ne). Ne is also found only as a gas and is also chemically inert. And eight more steps up gets us to Argon (Ar), also inert. Again we see the magic number 8. But then, ten steps further up gets us to the next inert gas, krypton (Kr), and ten more steps to the inert gas, xenon (Xe), and then, you guessed it, thirty-two more steps to radon (Rn), the heavy radioactive gas that seeps into our basements from below ground and is carcinogenic if inhaled. This set of elements forms another conspiracy, another gang whose members have similar chemical behavior, and we place these in another column in our periodic table, known as the "noble gases" (somehow "nobility" is linked to noninteractivity). These have to be gases because they are chemically inert and cannot bind to the atoms in rock. Radon is formed deep in the ground as a by-product of radioactive disintegration (mostly from radioactive decay of the element thorium, Th) and then slowly diffuses to the surface into our basements.[12]

So we can replace the police lineup with a table that consists of columns as lists of the various "gangs" and the gangsters in them. *This is a classification scheme*. Classification schemes are the first order business in any science where one must first classify things according to their properties, whether they are birds or worms, or insects or protein molecules, or stars or galaxies, or elementary particles. When we classify all the atoms that we find in nature, according to the similarities in their vast and diverse chemical properties, we obtain the periodic table of the elements (figure 25).

The atoms that have common chemical properties appear in columns. For example, the highly reactive, extremely poisonous, noxious gases that form strong acids in water define the column of "halogens." This contains fluorine (F) , chlorine (Cl), bromine (Br), iodine (I), and astatine (At). Very reactive metals, somewhat like the alkalis (but, e.g., not so explosive in water), define the column of "alkali earth metals," containing beryllium (Be), magnesium (Mg), calcium (Ca), strontium (Sr), barium (Ba), and radium (Ra). Then there are the noble gases,

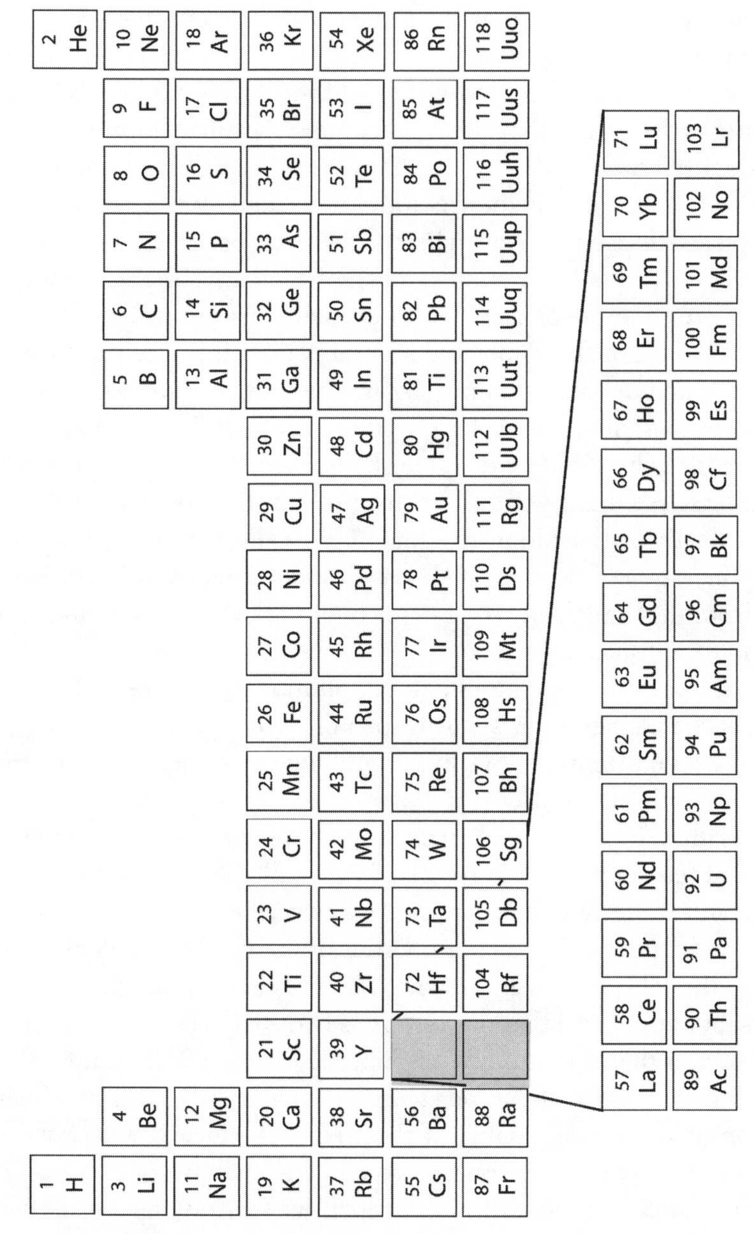

FIGURE 25: The periodic table of the elements. The periodicity and structure was explained by quantum theory (see figure 27 caption). Note that heavy atoms have unstable nuclei, and above uranium the elements are human-made, with some nuclei having ultrashort lifetimes.

helium (He), neon (Ne), argon (Ar), krypton (Kr), xenon (Xe), radon (Rn). And so forth.

Let's do a quick check to verify that the alkali earth metals have similar properties. Note that calcium, commonly found in milk, is readily absorbed (metabolized) into our bones and is the main bone structural component. If we metabolize too little calcium we become ill, for example, with osteoporosis, which causes our bones to become brittle. But we see from the periodic table of elements that a dangerously radioactive (nuclear physics again) element, strontium (Sr), is one step below calcium in the same column and is therefore chemically equivalent to calcium. Strontium, a by-product of nuclear explosions and prevalent in radioactive fallout, is therefore readily absorbed into our bones and is chemically indistinguishable from calcium. Once absorbed, the radioactive disintegration of strontium in the bones over time leads to the death of the blood-manufacturing cells that are located in the cores (marrow) of bones, which leads to leukemia. And if we draw further from that chemical column, we might ask if anyone would care for a barium milkshake?

One can get an enormous amount of information out of the periodic table of the elements. It is worth spending some time pondering it and thinking about the world we live in and how it is shaped by the properties of these atoms. The chemical periodicity as embodied in the periodic table is the first-order organizing principle for the understanding of chemical reactions. We've just seen an example of how it plays a key role in our metabolism and that toxic threats can be harbored by elements with similar properties. The properties of the rest of the elements comprising the table can easily be found by browsing some good chemistry books, but be assured that the pattern is faithful to the chemistry. For the most part, if there is a molecule composed of atoms Xab..c, then you can always make Yab..c if X and Y are elements appearing in the same column of the periodic table of the elements, since X and Y will then have identical chemical properties.

So what causes the remarkable repetitive (periodic) pattern of chemical properties to occur in nature? What internal structure would keep regenerating almost identical chemical properties in elements spaced eight steps apart, then ten steps, then thirty-two . . . ? Although chemists had developed the empirical rules describing the wildly divergent properties of the chemical elements, no one in the 1800s had an idea as to what was actually going on. The periodic table was a clue—in retrospect, a major clue—to the structure of chemical atoms.

HOW TO BUILD AN ATOM?

The chemical properties of a given atom are governed by the outermost electrons of the atom. These electrons are free to jump to and fro, from one atom to the next, and they can lead to the formation of molecules, the binding together of two or more atoms. This idea was a vague one, but it was "in the air" before the quantum theory arrived on the scene. With the more detailed understanding of the hydrogen atom, due to Rutherford, Bohr, and others, and the arrival of Schrödinger's equation, physicists could now launch a full-scale attack and try to understand the basic physical origin of chemistry.

The quantum theory successfully explains the chemical pattern seen in the periodic table of elements. In unraveling the pattern, many new and profound aspects of the quantum theory emerged. Schrödinger's equation could be applied to describe the motion of an electron around an atomic nucleus. He focused on the solutions to his equation that described "bound states," where the electron is attracted to the nucleus and binds with it to form an atom. Bound states are of immense importance in all of physics, and the nature of bound states, as we have seen from Bohr to Heisenberg, was the key to understanding the new laws of quantum theory themselves.

Bound states are most easily understood by considering a very simple example, an electron bound to a long molecule. It turns out that the form of the wave function of an electron that is trapped on a long molecule is exactly the same as the shape of the motion of a plucked string on a musical instrument, for example, a guitar string. In fact, we can easily work out the quantum energy levels of the trapped electron by thinking about the guitar string vibrations.

So get hold of a guitar (or other string instrument).

When you pluck the guitar string, it will vibrate, producing a beautiful musical note.

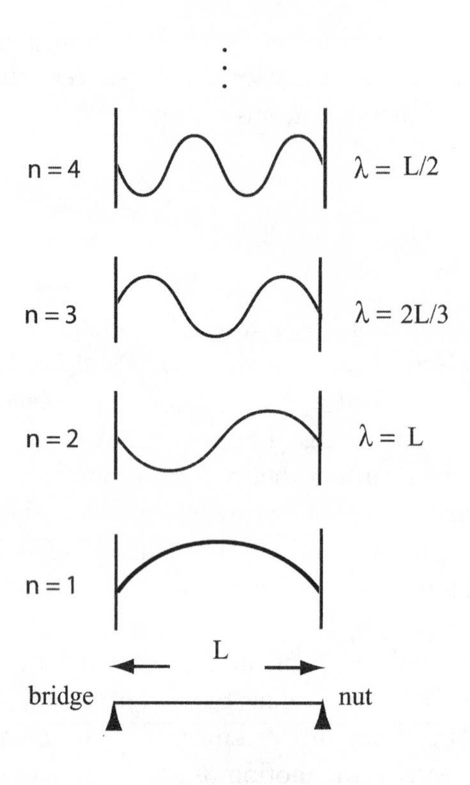

FIGURE 26: Waves on a musical instrument, such as a guitar string, are identical to the wave functions of the electron states (in a one-dimensional "potential energy well" such as in the molecule beta-carotene). The lowest mode, $n = 1$, is the ground state for an electron. The excited modes have $n = 2, 3, 4, \ldots$, and electrons can jump from one mode to another with the emission or absorption of a photon of the right energy (equal to the difference in the energies of the two states). Each mode can be occupied by one electron of spin "up" and one electron of spin "down."

Lightly pluck the string at its midpoint (preferably with your thumb, not with a pointed guitar pick). This excites the *lowest mode of vibration* of the string (because the pick is pointed, it tends to also excite higher modes). This corresponds to the lowest quantum energy state of motion of the electron trapped on a long straight molecule. This is the *lowest mode* or *lowest energy level* or the *ground state* of the system, corresponding to the lowest note of the plucked guitar string. The form of the wave is shown in figure 26.

The *second mode* of oscillation of the guitar string has a wavelength that is exactly half of that of the lowest mode. You can excite the second mode on a real guitar string with a little patience: hold one finger stationary at the midpoint of the string while plucking the string at about a quarter length of the string, then quickly remove the stationary finger. The stationary finger ensured that the center of the string didn't move when the string was plucked, which we see from figure 26 is a feature of the *second mode* of oscillation of the string. This special point where the wave motion is zero is called a *node* (not mode) of the wave function. The second mode has a pleasant and somewhat harplike angelic tone, one octave above the lowest mode. Because it has a shorter wavelength, the second mode for a quantum particle has a higher momentum and therefore a higher energy than the lowest mode.[13]

The next sequentially higher energy level is the third mode of vibration of the guitar string, which has one and a half full waves. This can be excited on the guitar by holding one's stationary finger at one-third the length of the string below the nut and plucking at the midpoint of the string, while quickly removing the stationary finger. One should then hear a very faint angelic fifth note (if the string is tuned to C, this note is G in the second octave above C). This corresponds in quantum theory to a still shorter wavelength for an electron wave function and a correspondingly larger momentum, and thus the energy is still larger.

There are real physical systems that behave exactly like this. In long organic molecules (molecules that involve carbon), such as beta-carotene—the molecule that produces the orange color of carrots—the electrons in the outer orbits of some carbon atoms become loose and move over the full length of the molecule, as if the electron was trapped in a long ditch. The molecule is many atomic diameters long but only one atomic diameter wide. This produces an electron wave function that is shaped very much like the modes of the guitar string. The photons that are given off by this molecule when the electrons hop from one quantum state (mode) to another have discrete energies that correspond to the difference between the two energy levels.

Even in the ground state, the electron is not at rest. It has a finite wavelength, therefore a finite momentum and energy. This ground state motion is called *zero-point motion*, and it occurs in all quantum systems. The electron in a hydrogen atom in the state of lowest energy is still moving. While it is not at rest, it cannot ever be in a state of lower energy. This is the lesson of quantum physics that stabilizes all atoms.

THE ATOMIC ORBITALS

By solving his equation, using brute force and mathematical dexterity, Schrödinger discovered a sequence of solutions for the modes of the electron motion in the simplest atom of all, hydrogen. The solutions are modes, like the modes of vibration of a musical instrument. Each mode for Schrödinger's hydrogen atom corresponds to a wave function, Ψ ("psi"), whose probability, Ψ^2, is a particular, fuzzy, cloudlike distribution for the probability of finding of the electron. Each mode has exactly the energy that Bohr had so successfully predicted with his "old quantum theory."

These modes, the distinct wave functions describing the electron, held to the atom by the attractive electric force of the nucleus, are called "orbitals." The orbitals have pretty much the same shape for any atom. Each electron in an atom moves in a particular orbital. The orbital shape, or distribution (Ψ^2), tells us where the electron can be found at any instant in time, only with—you guessed it—certain probabilities.

The orbitals for atomic hydrogen are shown in figure 27. They begin with the lowest mode, or the ground-state orbital, which is called the "1S state." While "S" should stand for "spherical," it actually stands for "sharp," from spectral nomenclature;[14] the 1 is called the "principal quantum number." The 1S state is perfectly spherical. When the electron is in this state, it has the greatest probability of being found close to the center, essentially on top of the nucleus. In fact, in the 1S state, the electron actually penetrates the nucleus.

But an electron can also be in other orbitals. Next highest in energy above the ground state we encounter the 2S and 2P levels. The 2S is also spherical, but it has a radial wavelike pattern, with a node separating the outer lobe from the inner one. There is zero probability of ever finding the electron at this node, but the electron can readily be found in either the inner or outer lobes. We also encounter the 2P states ("P" stands for principal, again following old spectroscopic nomenclature). In the 2P states, the electron can be viewed as revolving around the nucleus (while it is "breathing" around the nucleus in the S levels). There are three 2P states—$2P_x$, $2P_y$, and $2P_z$—that have dumbbell shapes when projected along three axes of space, (x, y, z). The 2Ps have the same energies (also, they happen to have the same energy as the 2S), and these are usually mixed quantum states; if we simply rotate the atom, the electron moves from one 2P orbital to another.

So a hydrogen atom, in its unexcited ground state, consists of a single electron orbiting the nucleus in a 1S orbital. The electron can be coaxed to jump to a higher energy state, like the next highest, or "2P"

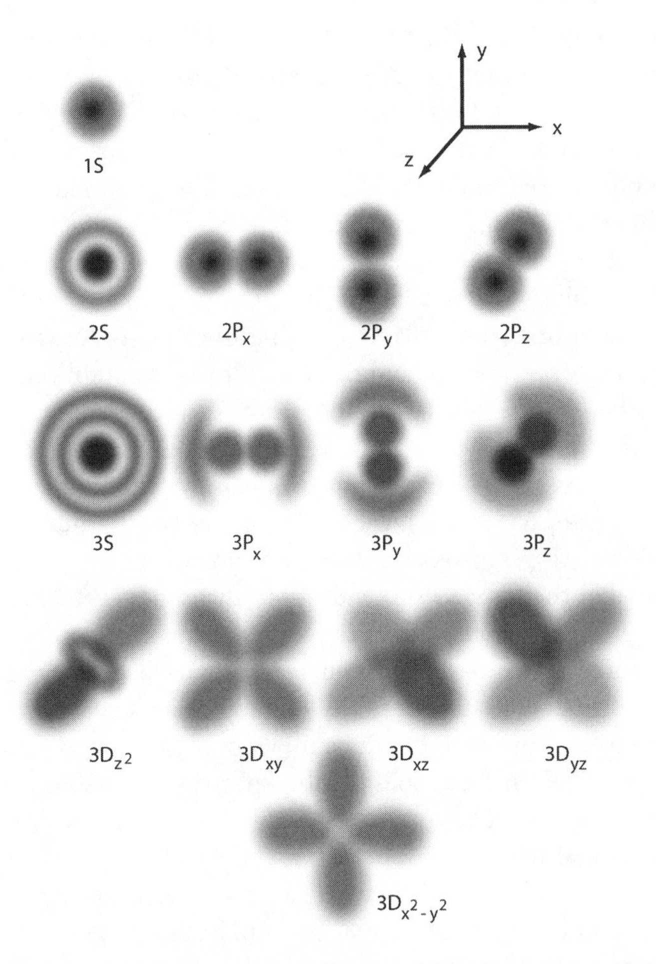

FIGURE 27: The lowest atomic orbitals. The hydrogen ground state is 1S, while helium has two electrons of opposite spins in the 1S. The second row of the periodic table (figure 25) represents the sequential filling of the 2S, $2P_x$, $2P_y$, and $2P_z$ states with at most two electrons each, one spin-up and the other spin-down. This explains the eightfold periodicity. The third row of the periodic table includes the filling of the 3S, $3P_x$,$3P_y$, and $3P_z$ states and is again eightfold periodic. The fourth row fills the 4S, $4P_x$, $4P_y$, and $4P_z$, and the five "3D" waves for a periodicity of 8 + 10 = 18. The fifth row fills the 5S, $5P_x$, $5P_y$, and $5P_z$, and the five "4D" waves for a periodicity of 8 + 10 = 18. In the sixth row we encounter the 4F states for the Lanthanides Z = 57 though 71, and the 5F states for the Actinides Z = 89 through 103. There is a high degree of complexity and mixing in the highest orbitals.

mode, or still higher, the "3D" states, by shining the appropriate energy photon on the atom, which the electron absorbs to gain the necessary energy to make the hop. After being in one of these higher modes for a short while, the electron will hop back down to the ground state, emitting a photon of exactly the correct energy as observed in the countless experiments that led to the development of the quantum theory (as described in earlier chapters). Solving Schrödinger's equation for the electron orbiting the nucleus of the hydrogen atom is much like solving the problem of a single planet orbiting the sun, as Newton did. This is called a *two-body problem*, and there aren't many complicating factors, so the mathematics is always fairly straightforward. One need only worry about the force between the nucleus (sun) and the electron (planet). When there are more planets in the solar system (as there are, Jupiter, Saturn, Venus, Mars, etc.), the math rapidly gets harder and exact solutions don't exist. This is also true for atoms with many electrons.

So what then is an atom with more than one electron? We begin by ignoring the electronic forces between various electrons and focus only on the force between the electron and the nucleus. Helium has a nucleus with charge $Z = 2$ and must therefore contain two electrons. Perhaps these two electrons could both be moving in the ground state, 1S orbital, together. This, in fact, does fit the spectrum of helium, but then we would seem to have a puzzle. Why is helium so much different chemically than hydrogen (recall, helium is chemically inert, while hydrogen is quite reactive)? Why isn't helium twice as reactive as hydrogen, having twice as many electrons? Why is helium, instead, chemically dead?

Next in the lineup comes lithium. So does lithium consist of three electrons piled on top of one another into the 1S orbital? Lithium, indeed, behaves like hydrogen, but helium behaves like neither. And what then of beryllium? Four electrons in the 1S ground state, yet different chemistry? What would account for Mendeleyev's bizarre periodicity if every atom simply consisted of Z electrons orbiting the nucleus piled into the 1S orbital?

Chemistry could not be explained by each atom having Z electrons piled into the ground state orbital, the 1S wave function. Every atom would then have essentially the same chemical properties, which is not at all what we have observed with our periodic table. And, if all electrons piled into one orbital, there would be no magic number 8, or magic number 10, which we discovered in the mysterious conspiracy of the atoms. Something more intriguing and more bizarre is happening here.

ENTER MR. PAULI, STAGE LEFT

Systems want to organize themselves to reach the lowest possible energy state. In atoms, the quantum rules, as in Schrödinger's equation, provide the possible orbitals—the allowed states of motion—that electrons can move in, with each orbital having its own energy level. The final breakthrough in understanding the atom is another remarkable and stunning realization: each orbital has room for only two electrons! If this were not so, the things of our world would be quite different.

This is where the great genius Wolfgang Pauli comes in. The irascible, legendary Pauli was the conscience of his generation who terrified his associates, who sometimes signed his letters "The Wrath of God," and about whom we'll hear much more (see chapter 5, note 1).

To prevent electrons from excessively overcrowding the 1S orbital, Pauli in 1925 proposed the "exclusion principle," which dictates that *no two electrons in an atom can be in the same quantum state simultaneously* . The exclusion principle controls the way heavier and heavier atoms fill up their electron orbitals and, incidentally, prevents us from walking through walls. Why? The electrons in your body are not allowed to occupy the same state as the electrons in the wall; they must be separated by lots of Wyoming-like great open spaces.

Professor Wolfgang Pauli was short, chubby, creative, and critical, with a sardonic wit that could be the delight or terror of his colleagues. As a teenager, he immodestly wrote a definitive article explaining the theory of relativity to physicists. And throughout his career he was prone to unforgettable one-liners that still reverberate in the community of physicists. Here is one: "Ach, so young and already you are unknown." Another: "That paper . . . isn't even wrong." And: "Your first formula is wrong. Nevertheless, the second formula doesn't follow from it." And: "I do not resent the fact that you think slowly, only that you publish faster than you think." It must have been a humbling experience to be on the receiving end of such observations.

Here's a poem about Pauli, by an anonymous author, as reprinted in George Gamow's book "*Thirty Years that Shook Physics*,"

> When with colleagues he debates
> All his body oscillates
> When a thesis he defends
> This vibration never ends
> Dazzling theories he unveils
> Bitten from his fingernails
>
> *Anonymous*[15]

Pauli's exclusion principle was one of his greatest achievements. It gave us chemistry by explaining why the periodic table of the elements is what it is. The Pauli exclusion principle simply states that in a given atom no two electrons can be in exactly the same quantum state. Verboten! And this simple rule allows us to build the elements in the periodic table and understand their chemical properties.

The buildup of elements in the periodic table follows two rules laid down by Pauli: (1) electrons must all be in different quantum states (the exclusion principle); and (2) the arrangement is designed to have the lowest possible energy. The latter rule also explains why objects fall under the force of gravity—for a body on the ground has less energy than one on the fourteenth floor. But we need two electrons to fit into the 1S orbital to make helium. Doesn't this violate Pauli's exclusion principle, which permits only one electron per quantum state? In fact, another great contribution of Pauli, perhaps his greatest, was the idea of "electron spin." (For more discussion of electron spin, see the appendix.)

Electrons spin—they are like little tops. They spin eternally and never stop. But each electron has two possible quantum states of spin: we call them "up" or "down." Therefore, we *can* have two electrons moving in the same orbital. This can respect Pauli's dictum "no two electrons in the same identical quantum state" because of electron spin: we place the two electrons into the same 1S orbital, but one electron is in the spin-up state, and the other is in the spin-down state. The quantum states are, therefore, not identical because the electrons have different spins. But, once we have done so, we are then finished loading up the 1S orbital with electrons. No third electron can join into this state.

This explains the atom helium. Helium has a completely filled 1S orbital, containing a pair of electrons. There is no room for another electron to fit into the 1S orbital of helium—the two electrons are snug as a pair of bugs in a rug. As a consequence, helium does not react chemically with other atoms—it is inert! Hydrogen, on the other hand, has only one electron in the 1S orbital and welcomes an additional electron to join it with opposite spin (another electron from another atom, hopping in to fill this orbital, is how hydrogen chemically binds with another atom, as we'll see shortly).[16] In the parlance of chemists, hydrogen has an "unfilled shell," a single electron in the 1S orbital, while helium has a "filled shell," two electrons with their spins in opposite directions in the 1S orbital ("shell" simply means "orbital," but it's an older term that you'll often hear chemists using). The chemistry is therefore as different as night and day between hydrogen and helium.

Now we're ready for lithium, with its three little electrons. Where do we put them? Two electrons go into the 1S orbital, spin-up and spin-down, and fill it like helium. Now, with the 1S orbital occupied, the third electron can only go into the next, lowest energy orbital—we have a choice as seen in figure 27, as four orbitals are available: the 2S, the $2P_x$, the $2P_y$, or the $2P_z$. The chemical properties of lithium depend only on this last electron since the 1S shell is now filled and this orbital itself becomes inert. The 2S orbital has very slightly less energy than the 2Ps, so the third electron goes into the 2S orbital. Thus, lithium is essentially identical chemically to hydrogen: hydrogen having a single electron in the 1S orbital and lithium having a single active electron in the 2S orbital. We are cracking the code of the periodic table of elements.

We can now start making the sequence of heavier atoms by filling up the 2S and 2P orbitals. For beryllium the next electron goes into one of the 2P orbitals (slightly repelled from the other electron in the 2S). The 2Ps are the dumbbell-shaped wave functions all having the same energy, and the electron can get into a quantum mixture of these states. Each 2S and 2P orbital can accommodate two electrons: one electron of spin-up and one of spin-down. Therefore, the sequence of simply adding more electrons into the 2S and 2P orbitals proceeds as we make heavier atoms, where the total number of negatively charged electrons balances the positive charge of the atom's nucleus in each case: beryllium ($Z = 4$), boron ($Z = 5$), carbon ($Z = 6$), nitrogen ($Z = 7$), oxygen($Z = 8$), fluorine ($Z = 9$), and neon (Z = 10). We get to neon by going eight steps up from helium. For the element neon, we see that each of the 2S and 2P orbitals is completely filled, having two electrons each, one spin-up and one spin-down. Just as helium consisted of a 1S orbital completely filled with two electrons, one up and one down, neon has a completely filled inner 1S orbital and also a completely filled 2S, $2P_x$, $2P_y$, and $2P_z$. We have discovered the origin of Mendeleyev's periodicity and the magic number 8.

So hydrogen and lithium are chemically identical because the hydrogen atom has one electron in 1S orbital, while lithium has one electron in its 2S orbital. Helium and neon are similar because He has two electrons completely filling the 1S orbital, while Ne has a completely filled inner 1S orbital and a completely filled 2S, $2P_x$, $2P_y$, and $2P_z$. Completely filled orbitals lead to stability and chemical inertness. Partially filled levels lead to chemical activity. The puzzle of the chemical properties conspiracy of our police lineup, as first discovered by Mendeleyev, has almost been completely understood.

But next comes sodium, loaded with eleven positive charges on the nucleus (Z = 11). Where do the eleven electrons go? Now we start with the 3S orbital and place a single electron there . . . voilà! Sodium and lithium and hydrogen must be chemically similar, each having a single unpaired electron in an outer S orbital. Then comes magnesium with an electron in a quantum mixture of $3P_x$, $3P_y$, and $3P_z$. As we continue higher, the filling of the 3S and 3P states is the same as for the 2S and 2P states, and again after eight steps, we get to completely filled orbitals and an inert gas, argon. Argon, like neon and helium, has a completely filled system of all orbitals, 1S, 2S, 2P, 3S, and 3P, each containing one spin-up and one spin-down electron. In making the atoms of the third row of the periodic table, we completely imitated what we did for the second row, as we filled S and P orbitals in the same way.

In the fourth row, however, the story changes. We start with 4S and 4P (these are not shown in figure 27) as in the second and third rows, but then we encounter the 3D orbitals (as shown in figure 27). These are still-higher solutions of Schrödinger's equation. The way these higher orbitals are filled up with electrons involves the details of having so many electrons present. Electrons interact among themselves through the electric force, and this leads to complications, something we have entirely ignored up to now. It would be like trying to solve Newton's equations for a solar system with many similar planets that are fairly close to one another in their orbits. Including all these effects is complicated and beyond the scope of our present discussion, but suffice it to say that it works. The 3D levels can mix with 4Ps, and so on, and we find that they accommodate a total of ten electrons before they are filled. Thus we have the pattern of "8" changing to "8 + 10," and so on. The physical basis of everyday matter, the basis of chemistry and, thus, how all things in biology are built, has now been established. The Mendeleyevian mystery has been solved.

MOLECULES

Now we are ready to build larger things—to make molecules. Pauli's exclusion principle, Schrödinger's equation, and the law that things are drawn into configurations of minimum energy tell us about the formation of molecules as well.

Molecules are combinations of two or more atoms that form a more complex bound state. We say that elements (atoms) combine together to form "compounds" (molecules). Here the combining atom's outer electrons do a new dance to form the new entity, the molecule (while the inner electrons in the filled orbitals do nothing). Again, we begin with the simplest physical system we can think of to begin an analysis.

We know that two hydrogen atoms form a hydrogen molecule, which we call H_2. The 1S orbitals of the individual hydrogen atoms, as we push the two atoms close together, gradually merge to form new orbitals with the two nuclei (protons) separated by their repulsive electric charges. This new molecular configuration represents a new set of solutions of Schrödinger's equation that occur when we analyze the electron wave function bound state attached to two nuclei (protons). The new ground-state orbital is called a σ *bond* (a "sigma bond") and it is the analogue of the 1S orbital. This is shown in figure 28, where the nuclei are represented by the two small dark centers within the electron cloud. There is also a higher-energy π bond ("pi bond") that is the analogue of the 2P orbital, as shown in figure 27. The σ bond, like the original 1S orbital, can only hold two electrons, provided their spins are different—up and down—the Pauli exclusion principle again at work.

The motion of the two electrons— as in the shape of the σ bond and the positions of the two nuclei within the molecule—are all determined by the minimizing of the total energy. This describes the simplest molecule, H_2, which is the form that hydrogen takes as a gas at standard room temperature and pressure. We often call the σ orbital, with two electrons, a *covalent bond.* Here the original two 1S states of the atoms are symmetrically merged together and the electrons are essentially cooperatively shared between the two atoms. You can think of the process in which two atoms get ever closer together and, if their outer electrons are just in the right orbitals, they can form the molecule. Molecule formation is therefore facilitated by the sharing of electrons, leading to more completely filled orbitals.

Yet another more extreme chemical bond forms between an atom like that of an alkali metal, hydrogen (or its partner sodium), which has a single electron in an outer S orbital, and an atom such as chlorine (a halogen). The halogen is one electron short of becoming inert, filling its outer orbitals. This leads to a very asymmetrical wave function for the donated electron as shown in figure 28. Here the electron from the alkali is donated, and essentially leaves the Na atom altogether to join up with the electrons in the Cl atom. The deserted Na atom now has effectively a lone positive charge and the Cl has picked up a net negative charge, so the pair of atoms is loosely electrically stuck together. This kind of bond—where an active outer electron deserts one atom and joins another, but the two atoms are still bonded by their opposite charges—is called an *ionic bond*. NaCl, ordinary table salt, is held together by an ionic bond. There is usually less energy released in forming an ionic bond (less binding energy) than in a covalent bond, so it's easy to dissolve salt in water where, as Na and Cl atoms separate and drift apart from one another. Batteries rely on this free mobility of ions (atoms that have lost, or gained, an electron to, or from, another atom in an ionic bond). The signal that is transmitted by nerve cells to produce sensations and thought involves an "ionic pump." It pumps atoms of sodium in and potassium out through the wall of the nerve fiber, controlled by the complex hopping of electrons among the atoms and their orbitals in the fatty cells that make up the nerve membrane. Life is a delicate balance of ionic and covalent chemical bonds.

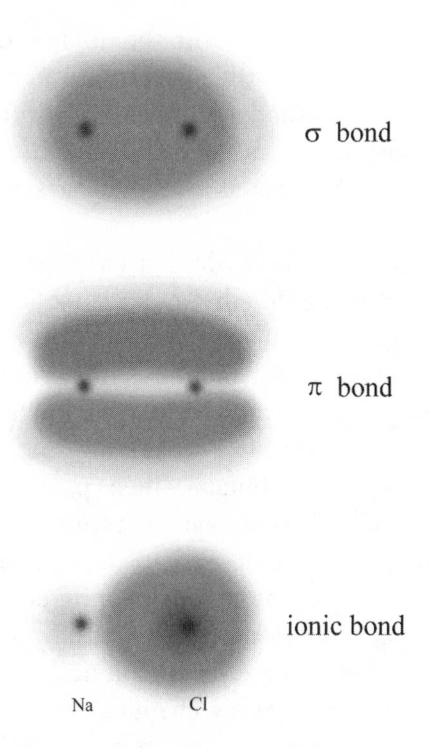

FIGURE 28: Simplest molecular orbitals. The σ bond is a covalent bond, such as in the H_2 molecule, and arises as the two 1S states are merged. The π bond is the merging of 2P orbitals. Covalent bonds occur when the electrons are almost completely tied to one atom as in NaCl (salt) where the Na electron jumps in to complete the outer shell of the Cl atom. The Na atom is then essentially an *ion* of positive charge attracted to the Cl *ion* of negative charge.

Let's consider the atom carbon, with Z = 6, which has two electrons in its 1S shell and four electrons occupying the 2S, $2P_x$, $2P_y$, and $2P_z$ shells. There's plenty of room here for four more "borrowed" electrons from nearby atoms to be added to the outer shells, at which point all the orbitals would be filled with equal numbers of spin-up and spin-down electrons. This makes carbon a very prodigious atom (if not the most prodigious) at entering into romantic, and fairly high-energy, covalent bonds with other atoms. For example, a carbon atom can easily form a molecule with four hydrogen atoms. The four outer carbon electrons are all paired into covalent σ orbitals, with the single hydrogen electrons and the resulting molecule taking the shape of a tetrahedron. This fundamental molecule, CH_4, is the basic molecule of organic chemistry (the chemistry of carbon) and is called *methane*. And there's an enormous amount of chemical energy stored in methane gas, waiting to be released by oxidation (combustion).

Now, we can remove one hydrogen atom from methane leaving a molecule, CH_3 , called the "methyl radical" or "methyl group," which has one electron eager to form a covalent bond. We can stick two methyl radicals together and get C_2H_6, an interesting structure (shown in figure 29) known as *ethane*. Now, remove one of the hydrogens (and you have the "ethyl group") and attach this to another methyl to form propane. Repeat the process to get butane, then pentane, hexane, heptane, octane—and so forth. This is just one pattern of construction of a large family of "macromolecules" that are known as "aliphatic hydrocarbons" (meaning "in-a-line" carbon atoms with hydrogen atoms). Clearly, with about one hundred atoms we can form a virtual infinity of different kinds of molecules, many of which are actually useful!

We noted above that there is a chemical reaction that methane participates in called rapid oxidation, or "burning": $CH_4 + 2O_2 \rightarrow CO_2 + 2H_2O$, or in words, one methane molecule meets two oxygen molecules and, with a spark of energy (a photon), rapidly combusts into one carbon dioxide molecule plus two water molecules. This reaction releases a lot of energy and is the basis of all carbon fuel burning. Note that burning of all hydrocarbons produces carbon dioxide as a by-product.

SUMMING IT ALL UP

We've seen that all the elements in the first column of the periodic table, our old friends the alkali metals, have one lone electron in the outermost S shell. The chemical activity of hydrogen (H), lithium (Li), sodium (Na), and so on, is the action, in molecule formation, of either sharing the electron with another atom (covalent bond) or getting rid of this outer electron altogether by donating it into the structure of another atom (ionic bond). So when two hydrogen atoms meet, they each contribute an electron to form a single filled covalent (σ bond) and merge into H_2, which has lower energy than two separated H atoms. On the other hand, a halogen like chlorine (Cl) needs just one electron to completely fill an outer shell. When Cl meets Na—ah, love at first sight!—chlorine grabs the loose sodium electron to fill its shell, and the extra negative charge causes the Cl to remain loosely bonded (an ionic bond) to the donor Na, and voila!—we've made sodium chloride: biblical salt—NaCl.

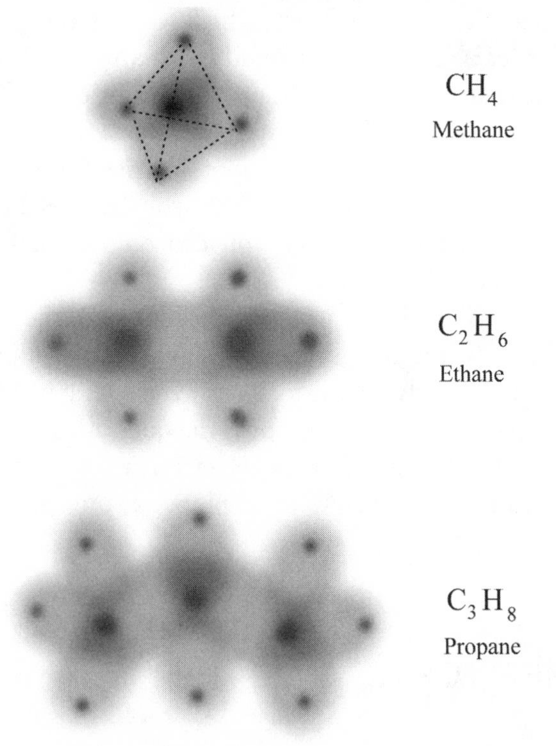

FIGURE 29: Chemistry in action. These are the three smallest members of the simplest family of hydrocarbons, called "aliphatic (in-line) hydrocarbons," methane, ethane, and propane. (The sequence continues with additional carbon atoms to butane, pentane, hexane, heptane, octane, nonane, decane, etc.)

We also know, emphatically, that two helium atoms do not combine. In fact, it is very difficult to get helium to react with anything. Why? Helium has a filled shell and therefore is standoffish and doesn't react chemically. Molecule formation is a "shell-filling" or "shell-completing" process, because filling a shell, even with a neighbor's electrons, generally makes a stable, lower-energy system. If you are borrowing the neighbor's electrons, you are, of course, capturing the neighbor in the process. Two hydrogen atoms combine so that their two electrons (one from each 1S orbital) fill a covalent σ bond and, behold, the hydrogen molecule! Oxygen is short by two electrons to fill its shell, so it attracts two hydrogens to make that stuff with which we gargle and take our daily baths. Get the idea? You are now a budding chemist.

We suspect that there are millions of you out there for whom the mere mention of chemical bonds may evoke a nostalgic trip back to high school chemistry and to an era of intense, oscillating joy and despair, never again matched and where you learned nothing about quantum theory. Fortunately, now you can contemplate such things without distraction.

Chemical reactions take place so as to minimize the energy of the combined system. In the quantum theory, the energy is quantized and electrons arrange themselves so as to occupy the lowest energy state. The exclusion principle takes precedence over the quest for the state of lowest energy, resulting in the rule that the electron will "place itself" in the lowest state that does not overlap or impinge on the space of the other electrons.

PAULI'S NEW FORCE

The strict prohibition against two electrons being in exactly the same identical quantum state leads ultimately to a prohibition against having any two electrons too close together. If the electrons—let us say both are spin-up—are squeezed down on top of one another, we would be forcing them, against the will of nature, into the same quantum state (the same place in space). So Pauli's rule *effectively* leads a resistance to this squeezing together, and in turn, this acts like a kind of force that is repulsive and keeps electrons apart (this is distinct from the electrical force between two equal sign charges, which is also repulsive but which arises from the electric field). This effective force generated by Pauli's exclusion principle is called the "exchange force." And this has consequences that go way beyond chemistry.

Pauli noted that the exchange force is an inescapable consequence of the probabilistic nature of quantum theory. The proof of its existence is a beautiful argument, worth reading twice (we give the "proof" in the appendix on spin, but it requires a more detailed mathematical background). For that matter, the Pauli exclusion principle arose as a consequence of the atomic experience: all electrons in the universe are exactly identical—not just very similar—but truly identical.

In our complex macroworld of ball bearings, sheep, and Supreme Court justices, such exact identity doesn't exist. Take a set of ball bearings coming straight from the ball bearing factory. They sure look alike, but are they identical? Microscopic examination will surely disclose some fine scratches or indentations that, if magnified, will give each ball bearing a different appearance. A series of precise weighings of four ball bearings may give 2.3297, 2.3295, 2.3299, and 2.3296 grams, as results, indicating very slight differences. Two cloned sheep or identical human twins will also be seen, at some level of inspection, to have great differences, reflecting their vastly greater molecular complexity. Not so for electrons.

All electrons have identical intrinsic properties. So what does this mean? If we swap any two electrons in their positions in an atom, we'll get back the same identical atom. There is no way, experimentally, or even in principle, to detect the swap. This is an example of *exchange symmetry*. But for the wave functions of the electrons in the atom, we find that when we swap any two electrons in this manner, we do get back the same overall wave function, with the two electron positions swapped, but it is now multiplied by –1. For electrons, this means that there is zero probability for the electrons to be in the same state at the same time. For example, if the electrons were at the same position in space, both having the same spin, and if we then swap two electrons, we would get a wave function that equals minus one times itself, and that could only be zero. (We explain this in greater mathematical detail in the appendix on spin).

The exchange "force" is strongly repulsive. However, it is not a real force like electricity or gravity—it is not associated with a "field." It is a consequence of the statistical probability being zero for two electrons to be in the same place at the same time. If a situation has a high probability of occurring, it would seem as if there were an a attractive "force" that makes it happen. Conversely, if the probability is low, it is as if a repelling "force" prevents it from happening. These so-called exchange forces are only an illusion, but they give us an intuitive sense of the effect.

Now, in the case of the Pauli exclusion principle, the wave function for two identical electrons to be in the same quantum state is zero—the ultimate repelling force. Such situations are excluded, as if there were a strong repulsive force. So now we can begin to see why we cannot sink our hand through a table or walk through solid walls, even though matter—the orbitals of atoms—is about 99 percent empty space. The electrons in your body cannot penetrate the "wall" atoms, because they are ruled by Pauli's exclusion principle, the prohibition against having electrons get too close together.

There's so much more to the fascinating details of chemistry, including the detailed chemical reactions and the complex molecules they lead to. Some of it is amenable to direct analysis by Schrödinger's equation, but much of it is far too complex for explicit calculations, and so much remains to be done. Quantum physics provides the foundations for chemistry. The extreme limit of complexity is now a major topic of interest to modern physics. How do we describe systems that are complex, and where do simple statistical models fail? We have no doubt that quantum physics explains it all, if only in principle. The devil is in the details.

Chapter 7

CONTROVERSY: EINSTEIN VS. BOHR… AND BELL

We survived the rigors of the previous chapters and have finally understood, thanks to Wolfgang Pauli, why there is chemistry (hence, biology) and why you'll probably never sink your hand through the granite kitchen countertop, despite all that empty space within it. Now we'll plunge ahead into deeper quantum mysteries and enter the great debate of Niels Bohr and Albert Einstein. Prepare to become intrigued. Let's start with a parable:

> Once there were four inveterate debaters who became hikers. Having trained together at MIT, they all retired within two years of one another, took up hiking, and continued their career-long discourse. They had learned that they could resolve their arguments only by a vote, and still remain friends. Yet, curiously, in all their heated arguments over the theory of everything, quantum technologies, where to put the next big particle accelerator, and so on, their vote was always *three against one*. Albert was always odd man out, valiantly defending a minority point of view. On one occasion when they hiked through Yellowstone National Park, Albert was once again alone, this time defending his claim that mathematical logic was always complete and that any mathematical theorem could be proved or disproved with enough effort.
>
> Despite an eloquent and impassioned exposition, he was outvoted again, as usual, *three against one*. But this time his convictions were so strong that he took an unprecedented step: he decided to appeal to the all powerful and beneficent, *Her*. Lifting his eyes upward, he intoned, "Please, o Lord, you know I am right! Give them a sign." Immediately, the cloudless sky darkened, and a purple-gray cloud descended over the four neophilosophers.

"See," said Albert. "A sign from *Her* that I am right!"

"Ach, nuts!" Werner answered, "We know that fog is a natural phenomenon."

Albert tried again. "Please give us a clearer sign that I am right!" The cloud turned suddenly into a vortex that rotated rapidly over the hikers.

"Again, a sign. I'm right! She knows and is telling us," shouted the excited Albert. "Well," said Niels, "I've seen such rotating cloud effects in Denmark. It's upper atmospheric turbulence." Max nodded in agreement. "Certainly no big deal."

Albert was insistent. "A clearer sign, *bitte*!"

Suddenly a piercing clap of thunder jolted the hikers and an awesome booming female voice descended from on high, screeching: "HE'S RIGHT!!!!"

Werner, Niels, and Max were shocked and conferred briefly among themselves, gesticulating and nodding. Finally with a resolved look on his face, Niels turned to Albert and said: "All right, so be it, we agree, She voted . . . so, this time it's three to two."

Scientific creativity is, ideally, a constant struggle between an intuitive direction and the need to have incontrovertible proof. Now we know that quantum science works over an awesome domain of the natural world. We've even found useful applications of it that have made the economy boom. We have also realized that the microworld—the world of the quantum—is weird . . . strange . . . bizarre. It's altogether unlike the science that evolved from the sixteenth century up to the 1900s. There has indeed been a true revolution.

From time to time, scientists, in conveying what they've found to the public, resort to metaphors, as we have. These are devised out of a degree of desperation to describe what we have seen, but in a "sensible" way, even to condition our own thinking to grasp a domain about which we have no everyday, hands-on experience. We surely don't have the language to describe this world. Our language evolved to accomplish other tasks. Alien observers, perhaps from the planet Zyzzx, who have only received data showing the behavior of large crowds of humans, would be aware of large parades, the World Series and the Superbowl, auto and horse races, the crowds in Times Square on New Year's Eve, armies on the march, occasional mobs attacking government buildings (in primitive countries, of course), and fleeing in panic from police actions. From a century of such data, the Zyzzxians would have a copious catalog of collective human behavior, but they would be totally ignorant of the capabilities and motivations of individual humans—their capacity for rational thought, for love of music or art, for sex, for creative insight, and for humor. All these individual properties would be averaged over, washed out in the collective behavior.

So it is in the microworld. When we are reminded that the hair on the eyelid of a flea contains a billion trillion atoms, we see why macroscopic objects—all the furniture of our human experience—prepare us not at all for the way nature orders the quantum world. Clearly, macroscopic nature blurs the properties of individual quantum objects—though not completely, as we will see later. So we have two worlds: the classical world, beautifully described by Newton and Maxwell, and the quantum world. Of course, in the final analysis there is only one world, the quantum world, and the quantum theory will successfully account for all quantum phenomena and will also reproduce the successes of classical theory. The equations of Newton and Maxwell will emerge as approximations of the quantum science equations. Let's reacquaint ourselves with some of the most shocking aspects of the latter.

FOUR SHOCKING THINGS

1. Our first challenge came with phenomena like radioactivity. Let's contemplate one of our favorite particles, the muon. The muon is a charged particle that weighs about two hundred times more than an electron, has the same electric charge as an electron, has apparently no size, that is, no radius, just like the electron (a pinpoint bit of matter), and spins on an axis, just like the electron. In fact, the muon, when first observed, was just an inexplicable heavy photocopy of the electron that caused I. I. Rabi to utter his famous exclamation, "Who ordered that?!"[1] But, the muon, unlike the electron, is unstable, that is, it is radioactive and it disintegrates, living only about two microseconds. More precisely, the half-life, or "average" lifetime, of a thousand muons is 2.2 microseconds (after 2.2 microseconds there are half as many muons as we started

with); however, we cannot predict exactly when any given muon (Hilda, Moe, Benito, or Julia) will decay. That event—the demise of Moe Muon—is undetermined, random, as if someone threw up a pair of dice that landed "snake eyes" every once in awhile. We must give up on a classical deterministic mechanism and replace it with probability as a foundation of fundamental physics.

2. There is, likewise, the conundrum of partial reflection, which you will recall from chapter 3. Light was known to be a wave—propagating, reflecting, diffracting, and interfering just like water waves—until Planck and Einstein discovered "quanta." Quanta are particles but still behave like they are waves. So a quantum of light, a "photon," heads for a glass window at Victoria's Secret. It is either reflected or it is transmitted through the window, either providing illumination of the suavely dressed mannequins in the window or a pale reflection of the guy on the street looking at them. We must describe this phenomenon by a wave function that is partially transmitted through the window and partially reflected from it, as waves do. But particles are discrete things—they are definitely transmitted through or definitely reflected. Therefore, our wave describes only the probabilities as to whether the photon will be reflected or transmitted. So we start with a wave of light coming toward the window from, say, the sun, Ψ_{sun}, which hits the window and is reflected or transmitted the way that waves do: $\Psi_{transmitted} + \Psi_{reflected}$. The probability that the quantum is reflected is $(\Psi_{reflected})^2$ and the probability that it is transmitted through the window is $(\Psi_{transmitted})^2$. These quantities are fractional, so they can only describe the chance that a whole particle goes through or is reflected.

3. Next is what we have previously dubbed the double-slit experiment (which harkens back to Thomas Young disproving Newton's "corpuscle" theory of light and replacing it with a wave). But electrons, muons, quarks, W-bosons, and so on, just like photons, are all described by waves. Thomas Young's experiment with light applies to all of them as well.

An electron can be emitted from a source and aimed at a screen with two slitlike holes, and so we can repeat Young's experiment with electrons instead of photons. An electron passes through the slits and is eventually detected far away on a detector screen. We repeat this experiment with electron detectors replacing photocells, counting one electron per hour, say, so we know that the electrons slip through the screen slits only one at a time (so there is no "interference" of one electron with another). As we discovered in chapter 4, when we repeat this experiment over and over and add up all the data as to where the individual electrons end up on the detector screen, a certain wavelike "interference" pattern appears. A single electron seems to know of the mere availability of two slits to travel through, but we do not know which of the two slits it went through, and this ambiguity leads, after counting up many electrons, to the interference pattern. The interference pattern changes completely if one of the slits is closed. It even changes if we also have a tiny detector that records which slit the electrons went through. We get the pattern of interference only when we are completely ignorant of which of two slits a single electron traversed. (Now stop and think about it: if this doesn't strike you as weird, then you should reread it.)

At the source, for any single electron, there is a "quantum wave" (a wave function) obeying the mathematics of waves, which travels through both slits and interferes as waves do. This results in the wave function at the detector screen, when both slits are open, to be the sum of two components, $\Psi_{slit\text{-}1} + \Psi_{slit\text{-}2}$. $\Psi_{slit\text{-}1}$ is the wave function for the electron traversing slit 1, and $\Psi_{slit\text{-}2}$ is the wave function for the electron traversing slit 2. The probability of detecting the electron at some point P on the detector screen is the mathematical square of this wave function. A little freshman high school algebra reveals that this is:

$$\Psi^2_{slit\text{-}1} + \Psi^2_{slit\text{-}2} + 2\,\Psi_{slit\text{-}1}\,\Psi_{slit\text{-}2}$$

and this describes the pattern seen at our detector screen. What emerges on the detector screen at the various points P—after we repeat this experiment many times with many, many electrons—shows the characteristic interference pattern as in figure 17 (see page 106). There are areas with many electrons (maximum probability) and areas with few electrons (zero probability), alternating across the screen. What we are seeing is the effect, from the above formula, of the *interference term*, $2\Psi_{slit\text{-}1}\,\Psi_{slit\text{-}2}$. The other two parts of the probability, $\Psi^2_{slit\text{-}1} + \Psi^2_{slit\text{-}2}$ are always positive numbers, and they are dull and uninteresting. They give the effect we would observe if we had no interference at all. For example, if we did the experiment with only slit 1 open fifty thousand times, then we would get something like figure 18 (see page 107). This would be a patternless distribution, a pileup of electrons appearing on the detector screen as described by $\Psi^2_{slit\text{-}1}$ (and, similarly, we would get a pileup described by $\Psi^2_{slit\text{-}2}$ if we allowed only slit 2 to be open). But the alternating bands, the interference pattern, comes from the $2\Psi_{slit\text{-}1}\,\Psi_{slit\text{-}2}$ term. This is not necessarily a positive number, and it oscillates, positive and negative, as we move across the detector screen. This is what produces the bands of light-dark-light-dark on the detector screen. It is the spooky essence of quantum theory that a single electron, a particle, interferes as it traverses the slits. This confirms the idea of a quantum state in which a definite discrete particle is neither in one state nor another but is truly in a schizophrenic, mixed state $\Psi_{slit\text{-}1} + \Psi_{slit\text{-}2}$.

4. If that weren't enough, we have to deal with all the additional spooky properties of particles. For example, there is the quantum property of "spin." Perhaps its weirdest aspect is that the electron has a "fractional spin." We say the electron is spin-1/2 ; it has "angular momentum" of magnitude $\hbar/2$ (see the appendix). Moreover, an electron is always aligned along any direction we choose to measure it, with either value of the spin $+\hbar/2$ or $-\hbar/2$, or, in our scientific vernacular, "up" or "down."[2] Perhaps the most freaky aspect of this is that if we rotate an electron in space, that is, if we rotate its wave function, $\Psi_{electron}$, all the way through 360°, then the quantum wave function becomes $-\Psi_{electron}$, that is, it returns to minus itself (we have devoted a section in the appendix to explain this). This is not what happens to any classical thing that we have ever seen before.

For example, take a drum baton, such as carried by the drum major or cheerleaders at a football halftime show. The baton points in some direction. The cheerleader rotates the baton by 360°, and, voila, it comes back to precisely its original orientation in space. But not so the electron wave function. When the electron is rotated through 360°, it comes back to *minus* itself. We're not in Kansas anymore—but is this just mathematical verbiage? All we can measure is probability—the mathematical square of the wave function—so how can we ever determine if the minus sign is there or not? What do minus signs have to do with reality? Might this just be self-absorbed philosophers staring at their navels with public funds for their research?

Nein! (says Pauli) This fact of the minus sign implies that, for two identical electrons (all electrons are identical), we must have a joint quantum state such that if we exchange the electrons then the wave function flips sign: $\Psi(x, y) = -\Psi(y, x)$ (see appendix). This leads us to the Pauli exclusion principle, the "exchange force," and the filling of the orbitals for electrons in atoms that gives us the periodic table and all of chemistry (the reason hydrogen is chemically active, while helium is chemically inert, etc.). This simple fact implies the existence of stable matter, the electrical conduction properties of materials, the existence of neutron stars, the existence of antimatter, and about half of the $14 trillion US GDP. And if the particles are like the photon (spin-1), then the swapping of a pair makes $\Psi(x, y) = +\Psi(y, x)$, which gives us lasers and superconductors, and superfluids, and the list goes on and on. All this marvelous stuff comes from a surreal and bizarre subuniverse of the quantum world and the Alice in Wonderland facts about identical photons and electrons.

HOW CAN IT POSSIBLY BE SO WEIRD?

Let's return to 1. and an old friend, the muon. The muon is an elementary particle that is two hundred times heavier than the electron and it decays in two-millionths of a second into an electron and some neutrinos (other particles of nature). Despite these whimsical features of the muon, we hope one day at Fermilab to build a particle accelerator that accelerates muons.

The radioactive decay of a muon is fundamentally determined by quantum probability. Evidently, Newtonian *classical determinism* is out on the street awaiting the garbage pickup. Ah, but not everyone was willing to give up on such a beautiful thing as "classical determinism," the exact predictability of the outcomes of physical processes in classical physics. Among the many efforts to rescue classical deterministic ideas was the concept of "hidden variables."

Suppose that inside the muon there is a kind of hidden time bomb—a tiny mechanism with a little windup alarm clock and a tiny stick of dynamite that causes the muon to detonate, albeit randomly. This mechanism would, of course, have to be a submicroscopic Newtonian mechanical device, so small that it is hidden from our eyes at the current level of microscopy, using our best detecting instruments, but it would account for the detonation, the radioactive decay, of the muon. When the hands of a little clock hit twelve—poof!—there goes the muon. If, in the creation of muons—typically through collisions of other non-muon particles—the internal muon clocks are all set somehow randomly (perhaps having to do with the details of the hidden mechanisms in the muon creation process), we would replicate the apparently random decay processes that we observe. *Hidden variables* is the name we give to such little mechanisms, and hidden variables could enter importantly into any efforts to modify the quantum theory, that is, in order to dispense with this "nonsense" of probability. Yet, as we shall see, eighty years into the debate, this approach has failed. Most scientists have accepted the weird logic of the quantum theory.

GENEALOGY OF MIXED STATES

In a beam of many random electrons, where the electrons can have all possible spin directions, there is always an equal probability of finding a single electron with spin-up or spin-down along any direction in space we choose to look. "Up" and "down" can be determined by passing the electrons through a powerful magnet with a particular nonuniform shape of the field (see the Stern-Gerlach experiment).[3] If the electrons are detected on a screen after deflection by the magnet, we see only two clusters of electron spots, one made by the spin-up electrons deflected up, and one made by the spin-down electrons deflected down. If we rotate the Stern-Gerlach magnet by 45 degrees, we still get two clusters, but they are now aligned up and down relative to the 45-degree line. The measurement seems to force an electron into a specific condition, up or down, for any choice of orientation of the magnet. But only probability can tell us what happens to a given electron.

These and many other examples led quantum scientists to the conclusion that atomic particles need not have definite values for their quantum properties until they are measured. One member of Bohr's research group, Pascual Jordan, proposed that the measurement act not only disturbs the particle but actually forces it into one of its various distinct possibilities. "We ourselves produce the result of the measurement," he asserted.[4] For his part, Heisenberg, too, insisted that the quantum domain consists of a world of possibilities rather than one of facts. All this is encapsulated in the quantum wave function, which describes all we can say about a given particle. From the wave function you can get the probability of finding the particle in a given place. According to the "Copenhagen" (Bohr) orthodoxy, the particle really exists in a variety of states and places, a mixed quantum state, each observable possibility having its own probability of being observed. If you look anywhere, at all possibilities, and add up their associated probabilities, these will total to 100 percent. There is 100 percent probability of the particle being somewhere!

The act of measurement therefore forces a system into a definite state and place at a given time. In mathematical language, the initial mixed wave function "collapses" into a precise state. For example, return to our photon as it reflects or transmits through the window at Victoria's Secret, which is now in the mixed state: $\Psi_{transmitted} + \Psi_{reflected}$. We detect the photon, and if we find that it has reflected, then we have disturbed the state and the wave function has "collapsed" into the new state $\Psi_{reflected}$. We call this a "pure state." We have resolved the ambiguity of the mixed state and we have restructured the wave function through the act of measurement.

Critics raised their voices (and some still do abound). The critics noted that the observer seems to be ill defined and that he/she/it intrudes too much on the processes of nature. This Copenhagen philosophy permits (just barely) the concept that one doesn't have to consider that an electron has a well-defined existence until it is observed by some he/she/it. The contrast with the classical world becomes too much to bear—for believers in classical reality, anyway. In 1935 Albert Einstein launched his counterattack, the most notable assault on quantum science ever. We'll get to that in a moment.

Meanwhile, bear in mind that we have a quantum theory that actually works, that makes predictions and explains phenomena that otherwise would not be understood. As Heisenberg insisted, quantum mechanics provides a consistent mathematical procedure that tells us everything that can be measured. So what is the problem? Some, like Einstein, hated the probability interpretation, the uncertainty relations, and especially the notion that one could not, even in principle, maintain (for example) that the electron has a well-defined trajectory from gun to screen. The idea that the electron is schizophrenic and that it somehow, in a confused mixed way, takes two independent paths to one destination makes no sense. Bohr's defense was that if you can't possibly measure it, it is meaningless to assume it takes a definite path. Others loathed the notion of wave-particle dichotomy. Heisenberg said, in effect, "They are all particles. The Schrödinger wave equation is just a calculational device; don't confuse those waves with the particles they describe. In essence we are dealing with something that is new and unprecedented in human thought and awareness . . . something that is neither particle nor wave, yet both at the same time—we are dealing with the 'quantum state.'"[5] Nature has revealed something profound and fundamental about herself that no one had conceived before the twentieth century.

> HORATIO:
> O day and night, but this is wondrous strange!
>
> HAMLET:
> And therefore as a stranger give it welcome.
> There are more things in heaven and earth, Horatio,
> Than are dreamt of in your philosophy.
>
> William Shakespeare, *Hamlet*[6]

In the period when quantum mechanics was racking up success after success—first, in the field of atomic science and next, from 1925 to 1950, in nuclear science and solid-state physics—the interpretation of quantum mechanics gradually jelled into two distinct schools of thought as the full implications of quantum science came into view. The quantum mechanics advocates, led by Bohr, counted among its leaders Werner Heisenberg, Wolfgang Pauli, Max Born, and nearly every other scientist living in or even passing through Copenhagen. On the other side of the ring were the skeptics and unbelievers, led by Einstein and Schrödinger and supported by other founders of quantum mechanics such as de Broglie and Planck.[7]

No one doubted quantum theory's successes, which were so great that some physicists began calling chemistry and biology, at their deepest level, "mere" branches of physics. The problem was the core probabilistic interpretation, a sharp departure from "classical realism," that is, from a belief that, even at the atomic level, objects must still exist with real, well-defined properties whether or not they are observed.

The Bohr-Einstein debate, often characterized as a battle for the soul of quantum physics, raged from about 1925 until the deaths of both men more than thirty years later. A new generation of physicists took up the campaign, which remains active today. However, most working physicists went their own way, exuberantly applying Schrödinger's equation, with its probabilistic wave functions, to all sorts of problems.

> Leon: Here I must insert my personal experience as one of the toiling masses who use quantum mechanics to earn his daily bread. In general, we experimentalists don't have so many chances to actually use Schrödinger's equation to compute something because we're too busy building electronic circuits, designing scintillation counters, and convincing committees to let us use the *#$@` ^&*$#% accelerator. But my Fermilab group had a unique opportunity once, in 1977, when in the course of our research we discovered an object never before seen. This new thing appearing on our screens could only be interpreted (we guessed) as a kind of "atom" consisting of a positive and a negative object, each of which weighed in at five times the mass of a proton. (It was called an "upsilon" or the Greek symbol ϒ). Inhaling the rich intellectual air of the 1970s, we reckoned that these unidentified objects were examples of a new bound state of a new quark with its new antiquark.
>
> At that time there were "up" and "down" quarks, "charm" and "strange" quarks, and rumors of a heavier quark family were in the air and even in the literature. Our new quark was named "bottom" (it had already been assigned a partner called "top," which wasn't observed for another twenty-five years!). We sometimes call it the "beauty" quark, or just the "b-quark." To obtain the properties of the "b-quark," one had to study how it and its antipartner behaved in the upsilon, the bound state of a pair, b-particle and b-antiparticle. To do that involved solving Schrödinger's equation (where the details of the force between the b-particle and the b-antiparticle, thought to be held together by a newly proposed force of "gluons," was not yet confirmed). Skeptics who were not convinced by our beautiful data called it the "Oops Leon," but eventually our discovery was confirmed. We were in a race with the theoretical physicists all over the world, people who were often desperate for something easy and new to calculate. (Theorists could solve

> Schrödinger's equation as easily as combing their beards.) We got to a reasonable answer first, but our calculations were soon outclassed by the herd of thundering theorists. However, we never stopped to contemplate if what we were doing using quantum physics to predict the behavior of the Upsilon was the right thing to do . . . of course it was!

HIDDEN THINGS

Back in the 1930s, long before anyone had knowingly met a quark, Einstein, deeply unhappy with the Bohr interpretation of quantum theory, embarked on a series of attempts to make the theory more like the good, old classical sensible physics of Newton and Maxwell. In 1935, aided by two younger theorists named Boris Podolsky and Nathan Rosen, he made his move.[8] As noted earlier, he proposed a thought experiment (a "*gedanken* experiment") to force a dramatic collision of logic between the quantum world of possibilities and the classical world of real objects with real properties, and to determine, once and for all, which was right!

This *gedanken* experiment, known to everyone as the "EPR paradox," after the initials of the authors, was designed to show that quantum science was incomplete. The authors hoped that a more complete theory existed and might one day be discovered.

What does it mean to be "complete" or "incomplete" if you're a theory? One type of "more complete theory" would involve the "hidden variables," mentioned earlier. Hidden variables are exactly what they say they are, unseen influences that affect the outcome of events that may (or may not) be revealed at a deeper level (like little internal time bombs that cause radioactive particles to decay). They are, in fact, a common feature of everyday life. If we toss a coin, the result is a head or a tail, with equal probability. This has been tried perhaps ten trillion times in recorded history, no doubt since the invention of the coin and certainly ever since Brutus tossed one to decide whether to kill Caesar. We all agree that the result is unpredictable—a result of a random process. But is it? Here is where hidden variables enter.

One such variable is the force used to flip the coin. How much of the force throws the coin up and how much causes it to spin around its diameter? Other variables are the coin's weight and size, the tiny air currents pushing and pulling on it, the exact angle at which it contacts the table surface when it finally lands, and how hard the table surface is. (Is the table made of hard slate or is it felt-covered?) There are, in short, a mess o' hidden variables influencing a coin toss.

Now let's imagine that we construct a machine that flips the coin precisely the same way each time. We'll use the identical coin on every try, shielding it from air currents (perhaps in the vacuum of a bell jar), and we'll insist that the coin always land near the center of the table's surface, where the elasticity controlling the bounce is exactly reproducible. After spending about $17,963.47 on all the apparatus, we're ready to try it: so let's go. The coin comes up heads each time! We toss it five hundred times. We get five hundred heads! We have managed to control all the sneaky hidden variables so that they are no longer hidden and no longer variable and we have defeated chance! Determinism now rules the day! Newtonian determinism applies to coins, arrows, artillery shells, baseballs, and planets. The apparent randomness of coin tossing, when described by an incomplete theory, was just the result of a large number of hidden variables, which could in principle be exposed and eventually controlled.

So where else is randomness acting in our everyday world? Actuarial tables predict roughly the longevity of people (or horses or dogs), but the theory of longevity of species is certainly incomplete, because there remain many complex hidden variables. They include genetic propensity to diseases, quality of the environment, nutrition, and likelihood of getting hit by an asteroid, among others. In some future world—barring occasional accidents—we might be able to greatly reduce the uncertainty about how long Grandma or Cousin Bob will live.

Physics has had some success with theories of hidden variables. Consider the perfect, or "ideal," gas law, which describes the relationship between the pressure, the temperature, and volume of a low-pressure gas confined to a container. Raise the temperature, and we find that the pressure goes up. Increase the volume, and the pressure goes down. All this is neatly described by a formula like PV = NRT ("pressure times volume equals the number of gas molecules times a constant R times the temperature"). However, in reality, there are bazillions of "hidden variable" here—the gas is composed of bazillions of molecules. Knowing this, we can define the temperature statistically as the average energy of a single molecule, the pressure as the average impact force of the speeding molecules as they hit a certain amount of area of the walls of the container, and N as the total number of molecules in the container. The gas law, formerly an incomplete description of a "gaseous medium," could then be completely and accurately explained by the "hidden" molecules and their average motion. In like manner, Einstein, in 1905, explained the jumpy motion of tiny grains of fine powder suspended in a jar of water (the so-called Brownian motion). This "random walk" phenomenon of the powder granule was an inexplicable puzzle until Einstein invoked the hidden process of impacts by the surrounding individual water molecules.

It was perhaps natural, therefore, for Einstein to suppose that the theory of quantum physics is in reality incomplete—that its apparent probabilistic nature is actually the result of averaging over hidden, still unseen, internal complexity. Suppose that one could reveal this concealed complexity—then we could go on to apply deterministic Newtonian physics to it all and restore an underlying classical reality. If, for example, the photons of light have internal, hidden mechanisms for preferring reflections or transmissions, then the randomness as they strike the window at Victoria's Secret is only apparent. If we knew the mechanism, we could predict definitively the reflection.

Let us hasten to assure you that no such properties have ever been discovered. Physicists, such as Einstein, were philosophically repelled by the idea of basic, fundamental, unpredictable randomness as a feature governing this world and they hoped that Newtonian determinism could be restored. If we know all the variables and can control them, we can organize the experiment so that the outcome is determined. This is the basic premise of classical determinism.

In contrast, quantum theory, as interpreted by Bohr and Heisenberg, holds that there are no internal variables, that the randomness and indeterminacy described by the uncertainty principle is intrinsic and fundamental to nature and exhibited in the microworld. If it is impossible to determine the exact outcome of any given experiment, it is impossible to predict the future course of events—determinism, as a philosophy of nature, is wrong.

The question is, can we know if there are hidden variables or not? Let's begin with the challenge.

THE EPR GAUNTLET: ENTANGLEMENT

Einstein, Podolsky, and Rosen knew what they had to do: reveal the incompleteness of quantum science. To begin, they obviously needed a clear definition of what it means for a theory to be complete. They said that a complete theory must contain all the elements of "physical reality." However, because quantum mechanics is intrinsically "fuzzy," for example, things can be in "mixed states" that are indefinite as to which of several possibilities they describe, the authors had to carefully define reality. They settled on the following reasonable condition: If, without in any way disturbing a system, we can predict with certainty (i.e., with probability equal to unity) the value of a physical quantity, then there exists an element of physical reality corresponding to this physical quantity.

If we are going to demolish a theory as useful as the quantum theory, we must carefully itemize our assumptions, so EPR followed up with a second assumption: the *locality principle.* This says that if two systems have been separated to such a distance so that communication between them in a finite amount of time is impossible (i.e., it would take signals traveling much faster than nature's speed limit, the speed of light, to communicate between them), then any measurement on one system cannot produce changes in the second distant one. This is commonsense, sweet reasoning, wouldn't you agree?

The EPR idea is to transmit two particles from one source to distant receivers. The measuring of the properties of one of the particles at one of the receivers can in no way influence the outcome of measurement at the other receiver. If it did, the locality principle would have been violated.

Well, there's a key point we have to get straight here. Let's say we have a friend who is the "source" living on the distant planet of Arcturus 4. Our friend owns two billiard balls, one red and one blue, and agrees to send to us one of the balls while sending the other to another mutual friend on Rigel 3. As soon as we open our package, to find a red ball, we know instantly that our mutual friend will receive a blue one. Does this violate locality? Of course not, we haven't in any way influenced the outcome on Rigel 3. Nothing has changed upon observing the contents of our package since classical states can never be mixed states: our friend, the source, knows who got which color, and we know there is definitely a billiard ball in the package of a definite color, as controlled by our distant friend who sent them.

But quantum theory is more bizarre. It allows for entangles states, that is, a state in which the source cannot know what it launched so that there is, for us, a certain possibility of receiving a red ball and a certain complementary possibility of receiving a blue ball. Which of these possibilities is beyond the control of the sender? This is, according to Bohr and Heisenberg, determined by the wave function that fills the whole universe and isn't known until it is measured. Once it is measured, the wave function collapses into one of the two definite possibilities. So, if we open our package and find a blue ball, with 50 percent chance, we have *caused* the probability at the distant star Rigel 3 to be 100 percent red. If we hadn't opened the package, the distant star had a 50 percent probability of observing red or blue. We, the interloping observer, seem to have caused something an infinite distance away to have instantaneously changed!

EPR then proposed a thought experiment to show that this violates the locality principle. In this experiment, a radioactive particle, at rest in the middle of an apparatus, decays into two equal-mass particles, which fly apart, east and west, with equal speeds. The source particle has spin-0, and when it decays it must produce a state in which the total angular momentum is zero. EPR assumes that it decays into two spin 1/2 particles, and these go flying out, back to back, one going east, the other heading west. One of the outbound particles going either east or west has spin-up ("up" is measured along a direction we'll call z that is perpendicular to the direction of motion) and the other particle going west or east has spin-down ("down" referred along the same z direction). This balance conserves angular momentum, which we know must hold in all realms of physics for other deep reasons (and which is known to be true in quantum theory).

The quantum state, however, is an entangled state, with a 50 percent chance of spin-up going east, with spin-down west; and an equal 50 percent, chance of spin-down going east, and spin-up west. This particular state is called an *entangled wave function* with two parts:

$$\Psi^{up}_{east}\ \Psi^{down}_{west} - \Psi^{down}_{east}\ \Psi^{up}_{east}$$

If we receive a spin-up particle (a spin-down particle) in Chicago in the west, then the other one traveling east to Beijing must be spin-down (spin-up). But the source does not know which of the two cases sent, other than sending us a mixed quantum state.

According to the Copenhagen interpretation, if we now measure the spin of one of the particles, say in Chicago, it will have one of two spins with 50 percent probability for each of "up" or "down." If we happen to measure spin-up, then the wave function collapses throughout the entire universe and we now have:

$$\Psi^{down}_{east}\ \Psi^{up}_{west}$$

(The minus sign doesn't matter now.) If instead we happen to measure spin-down in Chicago, the wave function collapses to be, again, throughout the entire universe:

$$\Psi^{up}_{east}\ \Psi^{down}_{west}$$

We have therefore caused a change in the wave function all the way over in Beijing—instantaneously—by measuring the spin in Chicago! In fact, even if the source was on Arcturus 4, and we repeat the same experiment, then detecting the electron on Earth changes the wave function on Rigel 3 instantaneously. This is clearly in violation of the locality principle as demanded by EPR.

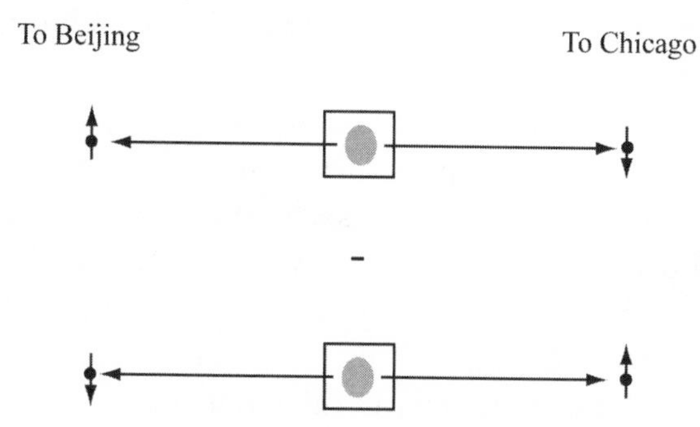

FIGURE 30: Einstein-Podolsky-Rosen considered an *entangled state* that is produced by the decay of a radioactive material that emits a pair of electrons, of the form:

(spin-up to Beijing, spin-down to Chicago) – (spin-down to Beijing, spin-up to Chicago)

(The minus sign is irrelevant; it could as well be +). The detection of a spin-up electron in Chicago instantaneously collapses the state into the form:

(spin-down to Beijing, spin-up to Chicago)

throughout the entire universe. Einstein thought this implied the instantaneous transmission of a signal through space and therefore a flaw in quantum theory.

By simply measuring the spin of one of the isolated particles, we have evidently "influenced" the spin of the other, now distant particle without touching it. Thus the second particle has a property that we have determined apart from being measured and apart from being there. Quantum theory denies that we can know something about an isolated particle without directly measuring it. So, concluded EPR, quantum theory is incomplete. (Recall that Bohr and his colleagues had accepted the notion that quantum science did not allow a particle to have any real physical properties unless it was measured, and this absence of "reality" in the absence of observation was part of what bothered Einstein.)

Let's look into our language more closely. A radioactive thing, at rest and with zero angular momentum, decays into two particles, A and B. The conservation laws of angular momentum say that particles A and B, flying apart in opposite directions, must have opposite spins and momenta. But quantum theory doesn't require A to have any particular spin. In fact, it allows the spin to be undetermined until measured. A precise measurement of the spin of particle A, light-years away, determines that A has a definite spin state, and the angular momentum conservation then forces B to have a definite spin pointing in the opposite direction. Measuring A must determine the spin of particle B many light-years away. But there is no way that measuring A many light-years from B can instantaneously influence B.

In a kind of *reductio ad absurdum*, Einstein announced that the only way B could acquire properties that result from measuring and disturbing A is if we somehow could send a message to B (such as: "We are measuring A. Spin is up. You'd better find B has spin down"). Since A and B are far apart (in *gedanken* experiments they could be megaparsecs apart), this message would have to be sent at speeds way beyond the speed of light—superluminal speeds. Einstein summarily ruled out such "spooky actions at a distance," observing, "My instincts as a physicist bristle at this."[9] EPR concluded, in short, that B's properties cannot be changed by measurements of A—a belief known as "locality"—and therefore B must have had the definite properties before the measurement. Since quantum mechanics can allow particles to have ill-defined properties like momentum or charge before they are measured, and since by measuring A we know the properties of B, Einstein then concludes that quantum mechanics is incomplete. There must be some deeper hidden variables.

Now, suppose (EPR continues) that we had precisely measured A's position rather than its spin. A has no definite position until we measure it since it is described by a wave function à la Schrödinger. Since, by the conservation of momentum, the two particles are moving apart with equal speeds but in opposite directions, at any instant they will have moved an equal distance from their source. Assuming locality by measuring A's position, we will instantaneously learn B's exact position. To repeat our argument, since measuring A cannot influence B (the locality principle), B must have had this exact position attribute before we measured A's position. Therefore B possesses a precise position and a precise momentum, simultaneously. You will no doubt instantly realize that this violates the Heisenberg uncertainty principle, the rule that a particle may not have precise values of position and momentum simultaneously. Do you see where we're going? Either quantum mechanics, embodying the uncertainty conditions, is incomplete, as EPR concludes, or B is affected by measurements at A. The latter would mean that nonlocal disturbances exist, and that, to EPR, is impossible.

Recall that Heisenberg's uncertainty principle states that if you try to measure the location of a particle, you inevitably disturb its momentum: the more precisely the location measurement is made, the more it disturbs the momentum (and vice versa). This is a very satisfying explanation of the uncertainty relation. Heisenberg teaches that we cannot know both the position and the momentum to arbitrary precision. As we've noted, the Bohr/Copenhagen school view is: if these can't be known, it is useless to believe they have these properties.

Einstein would argue in essence, "Okay, I agree we can never know the precise momentum and position of the particle simultaneously, but what is wrong with believing that it has these properties, well defined, simultaneously?" EPR had now challenged quantum physics by eliminating the disturbance factor and showing that we can know the momentum of B without touching it!

The impact of the EPR paper on Bohr was said to be "like a bolt from the blue." All traffic in Copenhagen reportedly stopped while the Master thought about it and discussed it with his colleagues.

When he had his answer, the issue became "trivial."

WHAT BOHR SAID TO EPR

The key to EPR is that our two diverging particles, A and B—having been born in the same act—are *entangled*, or *correlated* in their properties. A and B have indefinite values of momentum, position, spin, and so on, but whatever these are, they are entangled. If we measure A's exact velocity (momentum), we then know B's (exactly opposite); if we measure A's position at any time, we also then know B's. If we measure A's spin, then we know B's. When we measure, we change the wave function that previously allowed all possibilities for A's and B's properties. However, because of entanglement, when we learn about A in our laboratory here on Earth, we also learn all about B, which could be light-years away at Rigel 3, without poking, observing, or in any way disturbing it. We've indeed collapsed the myriad of possibilities for B down to definite ones instantaneously yet many megaparsecs away.

This does not refute the Heisenberg uncertainty relations in any practical way, because once we measured A's momentum, we disturb its position coordinate in an uncontrolled way. The idea is to argue that B must have precise values of momentum and position even if we cannot measure them. So what did Bohr finally decide? How did he respond?

After sweating it out for weeks, the Master Bohr finally concluded: "No problem!" Bohr insisted that the ability to predict B's velocity (via the measurement of A) does not mean that B has that velocity. Until you measure B, it makes no sense to assume it has any velocity. Similarly, B has no position until you set up a position measurement. What Bohr, joined subsequently by Pauli and the other quantum mechanics, was saying is: Alas, poor Einstein! He hasn't gotten rid of his classical obsession that all objects must have classical properties. In reality, you cannot know that B has certain properties until you measure and disturb it. Since you cannot know the property, it may as well not exist. Since we can't measure the number of angels that can dance on the head of a pin, they, too, may as well not exist. Nothing in practice violates the locality principle—you cannot use this to transmit an instantaneous anniversary card to Rigel 3 if you forgot your anniversary.

In one argument with Einstein, Bohr compared the quantum revolution to Einstein's own revolution, the relativity theory, wherein space and time took on strange new properties. Most physicists agreed, however, that the quantum view of the world is much more radical than relativity. Time, in relativity, is subjective.

Bohr reiterated: Once two particles are entangled in the microworld, they remain so, even if they are light-years apart. When you measure A, you are influencing the state in which both A and B reside. The spin of B, even if B is far, far away, is determined once you measure the spin of A. But that was Einstein's complaint. Bohr was noncommittal. This aspect of EPR is further exemplified by some deep insights from John Bell that came thirty years later. For now, the key word is "nonlocality," or in Einstein's admirable phrase, "spooky action at a distance."

In classical Newtonian physics, A("up") and B("down") is a totally separate and distinct state from A("down") and B("up"). The state is set up by the friend who mails the package. It is in principle knowable to anyone who looks at the data at the outset. The options are separate and totally independent, and opening our package simply reveals which is which. From the quantum theory point of view, the wave function describing A and B has the options entangled, and it is the wave function that changes instantaneously across space when one piece is measured—that's all—and still no observable signals can be sent faster than light—it's how nature works.

This authoritarian insistence may silence the newbie graduate student, but does it really help salve our philosophical souls? Certainly, Bohr's "refutation" did not satisfy Einstein and his team. The opponents were, in effect, talking past one another. Einstein believed in classical reality, that is, in the existence of well-defined properties of physical entities like electrons and photons. To Bohr, who had renounced the classical idea of an independent reality, Einstein's "proof" of incompleteness made no sense. It was Einstein's notion of reasonableness that was wrong. Einstein once asked a colleague of ours: "Do you really believe that the moon exists at that location only when you look at it?" If the same question is asked about an electron, the answer isn't so easy. The best one can come up with are quantum states and their probabilities. "Is the spin up or down?" For submicroscopic electrons coming off a hot tungsten wire, there is a 50–50 chance of either result, and, if no one is measuring, it makes no sense to say that the spin of a given electron points in any definite direction. So don't ask. Moons are much bigger than electrons.

A DEEPER THEORY?

Historically, whenever there have been two inconsistent theories in physics, there has been a search for a decisive experiment. In the Bohr-Einstein controversy, that was not so easy. Einstein had pointed out that none of the successes of quantum physics depended on these fundamental concepts and that, therefore, one should search for the deeper theory—"more sensible, less spooky"—that would still produce the Schrödinger equation and all the successful quantum results as a consequence. The Einstein group believed that additional elements of physical reality are associated with the particle, hidden behind the probabilities, and that quantum mechanics did not tell the whole story.

Bohr claimed that such a deeper theory *did not exist*, that quantum mechanics was complete, and if it was spooky, well, that was the way nature was. He and his crowd of radical revolutionaries were very comfortable with the notion that a particle is described by a set of probabilities as expressed by the wave function, and that is all that is necessary for a complete description of its physicality. Once a particle is measured, certain possibilities change to certainties with various probabilities. Bohr's group was not impressed (at least not publicly) by EPR and the way it cleverly avoided disturbing the particle whose properties were being determined at a great distance. The correlation in spin or position or momentum, wrought by the common origin of the A-B pair, permits us to learn all about particle B by measuring things about particle A, and our (secondhand) knowledge of B gives it the kind of reality that Einstein believed and insisted upon and Bohr did not. To Bohr, attributing a property such as spin, momentum, or position to a particle, when these are not measured, is "classical" thinking. Predicting that a measurement would yield a certain result does not mean that the electron actually possesses that property. Bohr really felt these properties should not be ascribed to the electron in the absence of a direct measurement.

> Leon: My two cents' worth: In case you are wondering what I think, I'll tell you. The experimental physicist in me wants to know what data distinguish the two points of view.
>
> The Einstein View: We agree we can't measure it, but it still has this spin or momentum.

> The Bohr View: If we can't measure it, it is meaningless to attribute this spin or momentum to it.
>
> These useless (to me) distinctions are dignified by purely semantic pizzazz. Does the B electron, unmeasured, possess independent elements of reality, as Einstein insists? Or is it "forced to adopt a particular value only through a measurement," as Bohr would have it? In my field of particle physics, we deduce the properties of particles in our collisions all the time. In a typical experiment we accelerate a proton and smash it into a second particle, producing perhaps fifty new particles going off every which way. The paths of the charged particles are recorded in our detector, while neutrals, such as neutrons, are detected later when they make collisions. Neutrinos, which can penetrate one hundred million miles of lead undetected, escape from our detector unscathed. We measure the momenta of all the outgoing charged particles and subtract this sum from the momentum of the incoming instigator of the collision. If there is a significant residue, we conclude that a neutral particle carried off some momentum. Accounting for the neutrons in this way, we learn a lot about neutrinos as well. In this application we make good use of the deduced momentum and thus favor the Einstein viewpoint.

There were probably deeper differences separating the two great theoretical physicists of the twentieth century, but it seems tragic that this issue apparently drove Einstein out of the physics mainstream and into the scientific isolation of his later years. The trouble was that Bohr, with his Copenhagen interpretation, was the ultimate arbiter of quantum mechanics, which was elegantly successful and used by the bread-and-butter physicists and chemists, whereas Einstein was the skeptic, continually raising questions that most of us didn't have the time or the intellectual weight to pursue. Einstein's questions made some of us uncomfortable. To doubt quantum mechanics was to seemingly threaten the very pursuits of our field, if not our sanity.

Of course, there were still puzzles that would penetrate the heavy scar tissue around our intuitions. The issue of the measuring apparatus itself was one of them. Aren't they made of atoms and other denizens of the spook-world? At what point does the classical reality typical of our world set in? A hundred atoms? A million? Do the quantum laws break down at macroscopic levels? Who or what kills Schrödinger's darn cat?

Who or what is an observer and gets to collapse a wave function throughout the entire universe?

Then there is the history: We all learned Newtonian physics and the long scientific voyage that established the reality of our world—the world of baseballs, moons, planets, bridges, and skyscrapers, extended by Maxwell's equations and applied to solar systems and the vast cosmos. All this supposedly now rests on an atomic world that does not recognize reality. Of course, theorists continued to try new ways of restoring reality, and their efforts littered the literature: multiple universes, hidden variables, nonlocal reality, and so on.

JOHN BELL

Then came a kind of denouement in the guise of a quiet-spoken Irish theorist working at the European Center for Nuclear Research (CERN).

> Leon: I'd been doing experiments at this lab in Geneva, Switzerland, since 1958. The facilities are great, the food splendid, and the nearby skiing beats anything in Batavia, Illinois. It was there that I met this young physicist with flaming Celtic red hair and piercing blue eyes; we had exchanged funny stories, and I knew he was working on the fundamentals of quantum physics, even though his job was to understand particle accelerators. This interest in abstraction was in sharp contrast to the bustling, very formal European lab, yet Bell was no navel-gazer. He was keenly observant, an expert in the arcane aspects of probability and statistics, and capable of doing highly technical, experimental, and theoretical calculations.
>
> CERN was founded in the early 1950s, partly at the suggestion of I. I. Rabi, my mentor at Columbia University. Rabi was a postwar mover and shaker in physics and in science policy. (It was he who suggested to President Eisenhower that presidents needed a science adviser.) It would take a collaboration of all the European nations to compete with the United States in particle physics, Rabi argued. Competition was lively, and CERN was intent on doing frontline research. That Americans were guests just illustrates the competitive collaborations that describe the field.
>
> Then, in 1964, John Bell found an experimental way of resolving

the problem of whether or not there were hidden variables in quantum physics. In particular, is quantum theory amenable to a complete, classical, deterministic über-description using local hidden variables? He discovered that in the back-to-back emission of two particles, as in the EPR thought experiment, there were certain directly observable statistical correlations that could, in principle, be measured and that could test Einstein's idea that particles must have real intrinsic (classical) properties. There are many ways to set this up, and they all fall under the rubric of "Bell's theorem."

BELL'S THEOREM UNVEILED

Bell's theorem would test a whole class of efforts to restore determinism via hidden variables and thereby support Einstein's classical reality beliefs. Bell devised a set of "inequalities": statements where X must be greater than or equal to Y, that were "obviously" true in classical systems and that could only be violated if entangled states were present with spooky actions at a distance. In the 1960s, when Bell proposed the "Bell inequalities," actual experiments to test them were unimaginably difficult, so they were just thought experiments, but by the late 1970s improved technologies enabled several groups of experimentalists to carry them out.

The short answer was clear: the theories claiming that particles had well-defined, classical properties were wrong. On the contrary, Bell's inequality violations were exactly as predicted by quantum theory. Quantum science was dramatically correct, utterly confounding classical reasoning. The particles were indeed described by probabilistic wave functions that could correlate over great distances and that must collapse instantaneously throughout all of space yet never violate the ultimate speed limit, the speed of light. Bell's theorem was a major breakthrough, graphically illustrating the profundity and counterintuitiveness of quantum theory and advancing our understanding as to what it is. It heightened our amazement at the extremely curious world of quantum reality.

It is worth describing Bell's theorem, but it's a bit of a mountain hike. (You can skip this section if you want; we've elected to present an even more detailed mathematical discussion in an "e-appendix" that is available as a downloadable pdf file at our website, http://www.emmynoether.com.) This follows on the development of spin, so it does require some immersion in some mathematical aspects of quantum theory. It doesn't involve very difficult mathematics but does demand some patience. But here in the text we can (almost) cast it all into English. Drum roll, please:

Einstein, in EPR, had concluded that quantum mechanics was incomplete. Although he did not propose anything new, he affirmed his belief that a complete theory based on classical reality, containing particles with well-defined properties, would eventually emerge. Whatever this new theory would be, it would have to conform to two basic principles. (1) Reality: particles exist and have definite physical properties; (2) Locality: if two systems are separated from each other for a reasonable time and distance, a measurement on the first system can produce no real changes in the second system. EPR needed this second principle because measurements on X can tell us properties of Y only if locality holds. If measuring X causes properties of Y to change, the EPR argument doesn't work. Of course, locality was intuitively favored over "spooky action at a distance." The new theory would have to involve quantum mechanics because quantum mechanics works.

Some efforts at such a theory had invoked hidden variables—unknown properties of particles that could generate probabilistic results in agreement with the experimental tests of quantum science. For example, suppose every radioactive particle has a hidden clock that determines precisely when the particle would decay. It would thus be a fundamentally deterministic system. However, the ensemble of such particles would permit one to calculate only the observed average lifetime of the particles and no more than a probability for when a given particle would decay.

Examining the EPR argument, John Bell—influenced by the theoretical physicist David Bohm and the subsequent arguments of Einstein—devised an experimental arrangement. His arrangement could distinguish between (1) the entire class of theories in which classical physics was used to simulate quantum mechanics, and (2) true quantum mechanics, with its intrinsic uncertainties and its denial of the existence of such properties as position or momentum between measurements.

The setup proposed by Bell was EPR-like. Electrons are emitted from a source and fly in opposite directions, into two detectors, 1 and 2, as in figure 30. Before leaving the source, the two electrons are in an "entangled" state, for example, in that their spins must add up to zero (this is a simplifying assumption; any total spin value will do, but the requirement that the total net spin of the two particles be a definite value is what enforces the relevant entanglement for us). Yet no particular value of an individual particle's spin is preferred. Another way to say this is that one electron can be spin "up" where "up" is defined as the direction perpendicular to the direction of motion (for simplicity sake). Then the other electron will necessarily be spin "down" relative to the same direction.

Our detectors are special. They are designed with a dial that allows us to measure the spin of an electron, either "up" or "down," in one of three chosen directions relative to the vertical (see figure 31). These spin measuring directions are "A position" $\theta = 0$ (vertical), or "B position" angle, $\theta = 10^0$, or "C position" a different angle $\theta = 20^0$ where θ is the angle from vertical. The detectors are set initially in some chosen way, say detector 1 in the A position, detector 2 in the B position. We then run the experiment for, let's say, a million radioactive disintegrations. We count, for example, the number of spin "up" electrons in detector 1 and the number of spin "down" electrons in detector 2. Then we repeat the experiment with different settings, perhaps with detector 2 in the B position, detector 2 in the C position, and so on. We repeat the experiment with all possible combinations of settings and count the number of events for millions of decays of the radioactive source for each setup. We then compare the results for each of the settings. What could be simpler?

John Bell had succeeded in deriving a "commonsense" prediction as to what would be expected if nature followed any one of the entire class of hidden-variable theories that obey the locality principles, for example, that no signals pass back and forth between detectors 1 and 2, that everything about a particle detected at detector 1 is determined by the particle at detector 1, not some far-removed object. His starting assumptions about these "classical" theories were so innocuous that almost anyone unacquainted with all the paradoxes of quantum physics would surely bet a large sum of money on their being always true. Yet the predictions of quantum theory distinctly and graphically violate these commonsense assumptions (see our e-appendix at http://www.emmynoether.com).

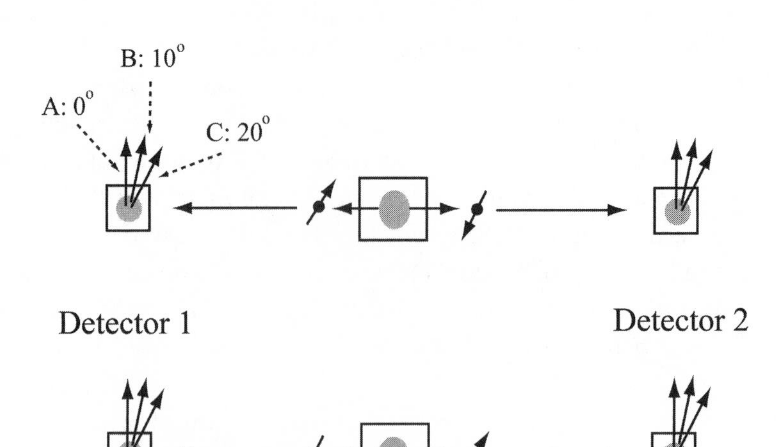

FIGURE 31: A Bell experiment. The entangled state is measured with detectors that can observe electron spins up or down relative to one of three angles, 0°, 10°, or 20°. Commonsense (classical) logic implies that when we run this experiment many times with different detector setting (the number of electrons that are spin-up in detector 1 and spin-up relative to 10° in detector 2), plus (the number of electrons that are spin-up relative to 10° in detector 1 and spin-up relative to 20° in detector 2) must exceed (the number of electrons that are spin-up in detector 1 and spin-up relative to 20° in detector 2). When we do the actual experiment, we violate this result. However, the result we get is consistent with quantum theory. Bell showed how quantum theory violates classical "commonsense" logic. The effect comes entirely from entanglement.

Leon: It is said that our educational system does not encourage its students to follow a reasoned argument. But I have every confidence that the reader will meet the challenge in this chapter, just as my mother did many years ago when she enrolled in an evening community college course in general physics. She had never finished high school and now at an advanced age was doing exceptionally well, having actually aced two quizzes. One day, after calling on her, the instructor said, "I read something in the *New York Times* about the Nobel Prize–winning physicist Leon Lederman. Are you related to him?"

"He is my son," Mom said proudly.
"Oh, that's why you are so good in physics!"
"No," Mom answered, "that's why he is so good."

BELL'S THOUGHT EXPERIMENT IN ENGLISH (ALMOST)

Let's look at Bell's thought experiment of 1965, which John Bell believed could be done someday (the actual experiments were performed in the late 1970s).[10] But first, let's take a trip to your local classical tropical aquarium.

We gaze into a tropical aquarium with large schools of several species of fish. We soon notice that every fish comes in one of two colors, either red or blue. From a logical point of view this is "binary," either a fish is "blue" or "not blue," and "not blue" is equivalent to "red," and "not red" is equivalent to blue (sort of like spin-up or spin-down). Soon we further notice that every fish, while either blue or red, is also either large or small. Later, we see that each fish has spots or no spots on it. So each fish actually has *three binary (two-valued) attributes*: (red or blue), (large or small), (spots or no spots). The "opposite of a big fish" is "not big" or a "small fish" and the opposite of a "spotted fish" is a "not-spotted" or "unspotted fish".

Now here is a deceptively simple but remarkably magical-seeming *theorem*. The following statement is true for *any* aquarium with any number of these kinds of fish, each having three attributes:[11]

> The number of red fish that are small, plus the number of big fish that are spotted is always greater than or equal to the number of red fish that are spotted.

Go ahead and read that back a few times. It's a little subtle, but quite simple, actually; you could show it to your classroom or friends, and perhaps even adapt it to a kind of parlor trick. (In the e-appendix we prove it, or see also note 8, but it is not hard to deduce in a "Sherlock Holmesian" way.)

Now let's introduce the notation of "N(X, not Y)," by which we mean "the number of objects in the group with attribute X and not Y." Therefore, the general statement we have made in English above is symbolically:

$$N(A;\ \text{not } B) + N(B;\ \text{not } C) \geq N(A;\ \text{not } C)$$

That is, in English, "the number of objects with attribute A but not B, plus the number of objects with attribute B but not C is always greater than or equal to the number of objects with attribute A but not C."

Following John Bell, we now adapt this logical statement in a "physically reasonable way" to our quantum mechanical experiment. We again consider the radioactive decay of some particle (source) that produces two outgoing identical particles, of opposite momenta and spins as in figure 31. The outgoing identical particles are traveling back-to-back, and the total spin of our outgoing particles adds to zero. This makes an entangled state. By the conservation of angular momentum, if the particle detected in detector 1 has spin-up, then the particle in detector 2 must be spin-down, and so forth.

If the outgoing states were actually "not entangled," as they are in classical physics, they would be of the form either (detector 1, up; detector 2, down), or (detector 2, up; detector 1, down) The conservation of angular momentum forces the total spin angular momentum of the state to be zero. This requires that the quantum state be entangled, and it takes the form:

(detector 1, up; detector 2, down) – (detector 2, up; detector 1, down)

(The minus sign here is actually associated with the net spin angular momentum of the pair of particles, which we take to be 0; this is not necessary, but it is the easiest assumption to make.)

Now, recall what our detectors do. They can measure the spins relative to any one of three possible angles from the vertical: "position A" with angle $\theta = 0$ (vertical); "position B" angle, $\theta = 10^0$, "position C " angle $\theta = 20^0$. So the list of the logical possibilities for the spin measurements we can get for either detector for the three different presets is

"A" = spin-up, along $\theta = 0$,
"Not A" = spin-down, along $\theta = 0$
"B" = spin-up, relative to the direction $\theta = 10^0$ direction
"Not B" = spin-down, relative to the $\theta = 10^0$ direction
"C" = spin-up relative to the $\theta = 20^0$ direction
"Not C" = spin-down relative to the direction $\theta = 20^0$

Note that we define "B" as spin-up in the B position, and "not B" as spin-down in the B position.

Suppose, for example, we set detector 1 to position A and set detector 2 to position A. Finally, we run the experiment. After collecting many events we would get a stream of data

Detector 1 (A): 1 1 –1 1 –1 –1 –1 1 1 –1 –1 –1 1 –1 . . .

Detector 2 (A): –1 –1 1 –1 1 1 1 –1 –1 1 1 1 –1 1 . . .

Each +1 is a spin-up measurement in the detector, and each –1 is a spin-down measurement. Notice that with each detector set to the A position (spin measured relative to vertical) there is a perfect correlation: if detector 1 measures spin-up, 1, then detector 2 must measure spin-down –1. This is just our old friend, the conservation of angular momentum. The result would be the same if both detectors are set to the B position or if both detectors are set to the C position. This result occurs true for either the entangled or unentangled case.

However, suppose we rerun the experiment with detector 1 in the A position and detector 2 in the B position. Now we might get the following slightly different data stream:

detector 1 (A): 1 –1 1 1 –1 1 –1 –1 1 –1 1 1 –1

detector 2 (B): –1 1 –1 1 1 1 1 –1 –1 1 1 –1 –1

Note that the correlation is not quite as perfect as before. Sometimes we get spin-up in detector 1 and also spin-up in detector 2. This happens because a pure spin-up state along the vertical is not a pure spin-up state along another axis that has been tilted by ten degrees. In slightly rotating detector 2 to position B, we are disturbing the quantum state of the particle when its spin is measured. However, locality would say that the disturbance should only affect the particle in the detector and should in no way be related to the particle in the other detector.

We run the experiment for a million radioactive decays and count the number of spin-up particles for detector 1 in the preset A position (angle vertical ($\theta = 0^0$)) and simultaneously the number of spin-up particles in detector 2 in the preset B position (for angle ($\theta = 10^0$)). Note that spin-up in detector 2 is the same as spin-down in detector 1 by the law of angular momentum conservation. So what we are actually measuring, referred to a common detector 1, is N(A, not B). We find, for example, with a million decays, a measurement of, say,

$$N(A;\text{ not }B) = N(1,\text{ spin-up }\theta = 0;\ 2,\text{ spin-up }\theta = 10^0) = 101\text{ events}$$

Then we repeat the experiment a million times and measure the number of spin-up particles for angle $\theta = 10^0$ at detector 1, simultaneous with spin-ups for detector angle $\theta = 20^0$ at detector 2. This is, again referred to a common detector 1,

$$N(B;\text{ not }C) = N(1,\text{ spin-up }\theta = 10^0,\ 2,\ \text{spin-up }\theta = 20^0) = 84\text{ events}$$

Then, we again repeat the experiment a million times measuring the number of spin-up particles for detector 1, angle vertical ($\theta = 0$) and simultaneously spin-ups for detector angle $\theta = 20^0$ at detector 2,

$$N(A;\text{ not }C\ ) = N(1,\text{ spin-up }\theta = 0^0;\ 2,\text{ spin-up }\theta = 20^0) = 372\text{ events}$$

Our simple logical hypothesis (a "Bell inequality")

$$N(A;\text{ not }B) + N(B;\text{ not }C) \geq N(A;\text{ not }C\ )$$

corresponds to

$$N(1,\text{ up }\theta = 0;\ 2,\text{ up }\theta = 10^0) + N(1,\text{ up }\theta = 10^0;\ 2,\text{ up }\theta = 20^0)$$
$$\geq N(1\text{ up }\theta = 0;\ 2,\text{ up }\theta = 20^0)$$

But what does the experiment tell us? Does quantum mechanics respect this simple logical hypothesis?

For the values of our experiment, we got: $101 + 84 = 185 \geq 372$. This apparently means that our Bell inequality is violated. Including statistical error for the finite numbers of signal events yields a discrepancy of

$$187 \pm 25\text{ events}$$

This means that our Bell inequality is violated with a high statistical significance (of about "4 sigma"). In statistical parlance, this is a pretty good measurement of a result, and we could do better with many more radioactive decays. It means that we can confidently say that quantum theory violates Bell's inequality.

What have we learned? Quantum physics does not respect the simple logic of classical aquarium fish. Entanglement is indeed the source of the violation of the inequality. However, even though the Bell inequality is violated, the quantum theory correctly predicts what is observed. (The exact results and the quantum calculation are presented in the e-appendix for those who want to continue along to the summit.)

Bell's theorem was published in 1964, almost thirty years after the EPR paper. All this time the EPR "paradox" had been discussed tirelessly by the wisest of the wise, yet no one had thought of looking at it in this way. It was just the kind of challenge to quantum mechanics that Einstein would have loved to devise. Of course, Bell had been able to review everything Einstein had written, as well as subsequent works by Born and by David Bohm, one of the authors of an alternative formulation of quantum theory.

Many versions of Bell's theorem were published in the decade following the original paper. Then, in a brisk period starting around 1979, came the experiments. The experiments confirmed the validity of quantum theory and the clear presence of entanglement. Given the vast number of successes of quantum theory, this is not a surprise. Yet the fact that simple classical logic now appears under assault is jolting. The results indicated to many physicists (and philosophers!) that there indeed existed nonlocal effects: measurements at detector 1 appear to cause an instantaneous change in detector 2. This is in violent disagreement with the predictions of classical reality and is at least compatible with quantum mechanics. But, in fact, quantum science calculations give the observed result. Bell proved that no classical hidden variable theory can, in principle, account for quantum behavior.

Subsidiary arguments extended the Bell theorem to rule out any (local) hidden variables that could account for the probabilistic nature of quantum theory while maintaining a deterministic, classical system. Follow-up experiments verified that the spooky "communication" between 1 and 2 occurs only when things are actually being measured, rather than, say, only when the measuring apparatus is set up.

This nonlocal, instantaneous communication does not violate the special theory of relativity. That theory forbids the transmission of information at speeds exceeding the speed of light. Analysis indicates that the spooky communication between 1 and 2 does not transmit information that an observer at 2 can use. Thus the Bell's inequality experiment reinforces the peaceful coexistence between quantum science and relativity.

Bohr would have been pleased. After all, his idea was that two systems, having once interacted, were forever connected "in an entangled system." It would be an error, he maintained, to think of A and B as having separate existences, even if they are vast distances apart. Bell seems to have proven him right.

NONLOCALITY AND HIDDEN VARIABLES

In short, the ingenious Bell was the first to design a crucial experiment to distinguish between the predictions of quantum theory on the one hand, and a class of theories that sought to reinstall "classical determinism" on the other. (As mentioned, many of the latter attempts used hidden variables and only looked probabilistic when one averaged over the hidden variables.)

Bell's real breakthrough was in the one element that concerned Einstein most: the spooky-action-at-a-distance idea, decried in the EPR paper. If A and B are connected so that any measurement of A influences B's trajectory, this fact "explains" EPR in a way that Einstein (and most of the rest of us) found most disturbing. Bell suspected, and later proved, that in fact all hidden-variable theories—which are constructed so as to agree with quantum mechanics—must be nonlocal. And he conceived a realistic, doable experiment to distinguish true quantum science from all the hidden-variable theories that obey locality theory and simulate the results of quantum science. Bell's theorem addresses the EPR paradox by establishing that measurements on object **a** actually do have some kind of instant effect on the measurement at **b**, even though the two are very far apart. It distinguishes this shocking interpretation from a more commonplace one in which only our knowledge of the state of **b** changes. This has a direct bearing on the meaning of the wave function and, from the consequences of Bell's theorem, experimentally establishes that the wave function completely defines the system in that a "collapse" is a real physical happening.

Say we put an electron in a box and then slide a wall in, dividing the box in two. We separate the two halves, but we don't know which one contains the electron. Now we transport one of the half-boxes to a station on the moon, leaving the other in New Jersey. The wave function gives us a probability of 50 percent that the electron is on the moon, and 50 percent that it is in New Jersey. Then we open the box in New Jersey and find the electron. The wave function collapses instantly to 100 percent probability—that is, yielding absolute certainty that the electron is in New Jersey. Without touching the box on the moon, we have changed its condition to "no electron." This "action at a distance," or nonlocality, is what Bell was seeking to establish experimentally—but without a moon station. The issue is whether the wave function is a real, physical quantity. If it is, as Einstein believed it had to be, then nonlocality is essential.

John Bell, who died in 1990, remained a modest man who often wore sweaters full of moth-holes even after his celebrity spread far beyond the physics world.[12] The baffling doctrine of nonlocality made him (and his theorem) well-known among many hip "New Age" types, who concluded that this was proof that everything was interconnected, sort of underpinning the urban cultural phrase "May the force be with you." Bell himself wasn't convinced of this: he was only sure that his theorem meant we didn't really understand what was going on. He once spent an enjoyable weekend at Maharishi International University in Iowa, where he very respectfully informed his hosts that his calculations didn't necessarily have anything to do with God.

WHAT KIND OF WORLD IS THIS, ANYWAY?

We have devoted this chapter to one of the most enigmatic aspects of quantum physics, the study of a new planet we call the microworld. If it were a new planet, subject to new laws of nature, that would be shocking enough, for it would undermine our understanding and control of all the scientifically based technology that makes us rich and powerful (some of us, anyway). But what is more unsettling is that the peculiar laws of nature in the microworld must cede to the stolid Newtonian laws when we get up to the level of baseballs and planets.

All the forces we know of—gravitation, electricity, strong/weak—are local forces. They get weaker as the objects exerting the forces separate; they propagate with a velocity sharply limited by the speed of light. Then along comes Mr. Bell, forcing us to consider a new, nonlocal influence that propagates instantaneously and weakens not at all with increasing distance. He begins with an assumption that such an influence does not exist, then goes through a series of logical steps to find his assumption contradicted by experiment.

Does this compel us to accept these otherworldly nonlocal actions at a distance? We are, indeed, in a philosophical quagmire. The deepening realization of how different the world is from our experience inevitably influences subtle changes in our thinking. The applied quantum science of the last eighty years has been a recapitulation of the vast successes generated earlier by Newtonian and Maxwellian physics, which defined the classical epoch. Surely now we have arrived at a deeper level, for quantum science underlies all our sciences (giving us classical physics as an approximation) and successfully describes the behavior of atoms, nuclei, and subnuclear particles (quarks and leptons), as well as molecules, the structure of solids, the birth pangs of our universe (quantum cosmology), the mega-molecules that define life, the current frenzy of biotechnology, even, perhaps, the operation of human consciousness. It has given us all this, yet philosophical and conceptual problems bedevil us, mixing unease with great expectations.

Leon: Something indescribably beautiful must emerge from all these terrors and wonders. It always has; it will again. (I think.) Artists create their visions of beauty out of the machinery of the imagination; the beauty of science lies in the sightings of nature's grace. You don't have to be a quantum mechanic to wonder at the night sky of winter far from city lights or (as I write here) the ragged peaks of the Tetons seen from the Idaho side. Although this year has been atypically hot and dry, never have the wild columbine, larkspur, fireweed, and lupine so strained nature's palette and never have the berries been so full and sweet. This face of nature can be perceived by all of us, but too few have been able to glimpse the invisible order in the world where quantum science reigns, enigmatic, beckoning us toward a final conquest or (possibly) toward hidden domains of endless vistas.

Fortunately, we physicists are a resilient breed. Few of us have actually had to be sedated after contemplating the strange land of John Bell.

Chapter 8

MODERN QUANTUM PHYSICS

In the previous chapters we have traced the struggles of the geniuses of the twentieth century who cobbled together quantum physics. On this journey we followed the development of the fundamental concepts that were radical and counterintuitive to those familiar with three hundred years of physics that was generated by Galileo and Newton. Some physicists, confronted with the quantum theory, had problems with the foundational issues, for example, the validity and limitations of the "Copenhagen interpretation" (and some still challenge and test them to this day). Most scientists, however, realizing they had a powerful new tool to understand the atomic and subatomic world moved on. They simply accepted and used the new physics, even if it didn't satisfy their philosophical tastes, and they created new subdisciplines of physics that continue to this day.

These have radically reshaped our lifestyles and our understanding, and especially our competence, within the universe. Next time you or a family member are lying in an MRI scanner at a hospital (we hope never), as it whirs, ratchets, taps, and boings in a loud, otherworldly orchestration, and a detailed image of your inner organs forms on a monitor in the control room, you are wholly immersed in the applied quantum world of superconductors, nuclear spin, semiconductors, quantum electrodynamics, quantum materials, chemistry, and more. You literally are an EPR experiment when you are in an MRI scanner. And if it's a PET scan the good doctor has ordered, you or your loved one are being dosed with antimatter!

Further progress beyond Copenhagen involved using the established rules of quantum theory to tackle many specific practical problems in new and previously intractable environments. Scientists now focused on what controls the behavior of materials. How do materials change phase, for example, from solid to liquid to gas, or to other kinds of phases, such as magnetized or demagnetized, when heated and cooled? What determines the electrical properties of materials; in other words, why are some materials insulators and others good conductors of electrical currents? These questions lie largely in the realm of "condensed matter physics." Most of these questions can be answered by using Schrödinger's equation—though new and sophisticated mathematical techniques have been developed along the way. The development of such new mathematical and conceptual tools is where sophisticated new gadgets came from, such as the transistor and the laser, and from these emerged the digital information technology world we inhabit today.

Most of the multitrillion-dollar economy—the part that derives from or is facilitated by quantum electronics and condensed matter physics—does not depend on Einstein's theory of special relativity but is simply "nonrelativistic," meaning they involve velocities much less than the speed of light. The Schrödinger equation is nonrelativistic and provides an accurate approximation of the world in which electrons and atoms are moving at velocities that are small compared to the speed of light. Assuming small velocities compared to light is a good approximation for the chemically active outer electrons in the atom, the electrons in chemical bonds, as well as the electrons moving in materials.[1]

However, many challenging questions remained, such as: What holds the nucleus of the atom together? What are the basic building blocks of nature, the elementary particles? How do we incorporate special relativity into the fold of quantum theory? Here we enter a world beyond the slowly moving systems that occur in materials. To address the physics of the nucleus, where mass can be converted to energy, as in a radioactive disintegration—nuclear fission or nuclear fusion—we indeed require an understanding of quantum physics when velocities are approaching the speed of light. This leads us into the hard-core realm of Einstein's special theory of relativity. And once we've understood its workings, we can follow with the more complicated and profound general theory of relativity (gravity). And the most fundamental problem of all, which remained unsolved until shortly after World War II: How do we describe in complete detail the interactions of a relativistic electron with light?

MARRYING QUANTUM PHYSICS TO SPECIAL RELATIVITY

Einstein's theory of special relativity is the correct formulation of relative motion in physics, including relative motion near the speed of light. It is a basic statement about the symmetries of the laws of physics.[2] It has a profound consequence on understanding the dynamics of all particles. Einstein derived the basic relationship between energy and momentum, a relationship that is profoundly different than that of Newton. This innovation in thought is what reshapes the behavior of quantum mechanics into its relativistic form.[3]

So, the natural question arises: What happens when special relativity marries the quantum theory? The answer: Something rather incredible.

$E = mc^2$

We have all seen the famous equation: $E = mc^2$. It is oft emblazoned on T-shirts, opening graphics for TV shows like the *Twilight Zone*, corporate logos, commercial products, and countless *New Yorker* cartoons. $E = mc^2$ has become a universal emblem for "smart" in our culture today.

We rarely, however, hear talking heads on TV correctly explain what this means. Usually they say something like "it means that mass is equivalent to energy." FALSE! Mass and energy *are actually two completely different things.* Photons, for example, have no mass, but they can and do have energy.

In fact, what $E = mc^2$ means is actually rather restrictive: in English, it literally means "a particle *at rest* that has mass m also contains an amount of energy E given by $E = mc^2$." So a heavy particle can, in principle, spontaneously transform (or "decay") into lighter particles and in the process yield up this energy.[4] This is why nuclear fission—the spontaneous "flying apart" of unstable heavy atomic nuclei into lighter nuclei, such as the fission of U^{235} (uranium-235)—produces a lot of energy. Likewise, light nuclei like deuterium can combine to make helium through a process called *nuclear fusion* with the release of a large amount of energy. This can happen because the helium nucleus has lower mass than the sum of the two deuterium nuclei. Therefore, energy can be released when two deuterium nuclei are squeezed together to make helium. These energy-conversion processes simply could not be understood before Einstein's theory of relativity and yet they are the reason the sun shines and life on Earth, including all its beauty and poetry, exists at all.

In fact, however, if an object is in motion, the famous formula $E = mc^2$ is modified.[5] Einstein knew this and he gave us the full formula (see note 3). The bottom line is that what Einstein *really* said, for a particle at rest (a particle with zero momentum) was *not* $E = mc^2$ but rather

$$E^2 = m^2c^4$$

This may seem to be an absurd distinction, but there is a big difference, as we'll see in a moment. To get the energy for the particle we have to take the *square root* of both sides of this equation, and sure enough, we'll then get a solution: $E = mc^2$. However, we get more!

Bear in mind a simple mathematical fact: *every number has two square roots*. For example, the number 4 has the two square roots, $\sqrt{4} = 2$ and $\sqrt{4} = -2$, the latter is negative 2. That is, we know that $2 \times 2 = 4$, but we also know that $(-2) \times (-2) = 4$ (two negatives make a positive when you multiply them together). The "other" square root of a positive number is a negative number. So we must keep and understand the two solutions: $E = mc^2$ and $E = -mc^2$.

So, then, here's the puzzle: How do we know that the energy we derive from Einstein's formula should be a positive number? Which square root is it? How does Nature know?

People didn't worry very much about this at first. It seemed like a stupid or "empty" question. Cocktail party sophisticates quipped, "Of course—everything has either zero or positive energy! What an absurd thing would be a particle with negative energy! Aren't we a bit insipid to even contemplate this?" They were too busy playing with Schrödinger's equation, which applies only to slowly moving electrons, as in atoms, molecules, and bulk materials. The question never arises in the nonrelativisitic Schrödinger equation, where the kinetic energies of moving particles are always positive numbers. Common sense would tell us that the total energy, especially the energy of a massive particle at rest, mc^2, must always be positive. Hence, physicists in the early days of special relativity simply refused to talk about the possibility of the negative square root, saying it must be "spurious" and that it "doesn't describe any physical particles."

But suppose such *negative energy particles* exist, particles in which we take the negative square root? These particles would have negative rest energy of $-mc^2$. If they were in motion, their negative energy would become a greater negative quantity, that is, they would *lose energy*, their energy becoming more and more negative if their *momentum was increased.*[6] They would continually *lose* energy by colliding with other particles, and by radiating photons, and their velocities would increase in the process (approaching the speed of light)! Their energy would become more and more negative, never stopping, and eventually the particles would have an infinitely negative energy. Such particles would fall into an abyss of negative infinite energy. The universe would eventually be full of these negative but infinite energy oddball particles, constantly radiating energy as they fell deeper and deeper into the infinite negative energy abyss.[7]

THE CENTURY OF THE SQUARE ROOT

It is rather remarkable that the entire thrust of physics in the twentieth century was essentially the problem of "taking the square root." When viewed in reverse, quantum physics was the problem of constructing "a theory of the square root of probability." The result was the wave function of Schrödinger, whose mathematical square is the probability of finding a particle at some place and time.

When we take square roots of ordinary numbers, weird things happen. For instance, we can get *imaginary* or *complex numbers*. And indeed, many weird things did happen in inventing quantum theory, when we did in fact encounter the infamous square root of minus one, $i = \sqrt{-1}$. Quantum theory necessarily involves i because of its inherent nature, based on square roots, and there's simply no dodging it. But we also encountered other weird things, like "entanglement" and "mixed states." These are also the "exceptional cases"—consequences of formulating a theory based on the square root of probability—which allow us to add together (or subtract) these square roots before we square the total. In doing so we can get cancellations, thus producing the phenomenon of interference, as seen in Young's experiment. These weird aspects of nature are perhaps just as counterintuitive as $i = \sqrt{-1}$ would have been to earlier civilizations, such as the Greeks. Recall that the Greeks initially even had trouble with irrational numbers: Pythagoras allegedly drowned a student who proved that $\sqrt{2}$ is irrational and cannot be written as the ratio of two integers. Though the Greeks finally accepted the irrationals by the time of Euclid, they never got to the imaginary numbers, at least to the best of our knowledge (see the "digression on numbers," chapter 5, note 5).

Yet another stunning physics result of the twentieth century, a consequence of this mathematics, was the concept of electron spin, described by the *spinor* (see appendix on spin). A spinor (pronounced "spin-or") is a square root of a vector. A vector, you'll recall, is like an arrow in space with both a direction and a length that can represent, for example, the velocity of an object. The square root of something that has a direction in space is a bizarre notion, with bizarre implications. When we rotate a spinor through 360° it comes back to minus itself. From this the mathematics dictates that, exchanging the positions of two identical spin-1/2 electrons, the wave function of the state containing two particles must change sign: $\Psi(x, y) = -\Psi(x, y)$. Pauli's exclusion principle follows from this fact: two identical spin-1/2 particles ("spin-1/2" means that their spin angular momentum is described by a "spinor") cannot be placed into the same identical state or the wave function would be zero. Recall, we can put two electrons in the same state of motion, for example, an orbital, if one has spin "up" and the other has spin "down," as in the helium atom, but then we're done, and the next electron has to go into another orbital; we can never put two electrons of spin up into the same orbital). Hence there is an effective "repulsive exchange force" between particles with spin-1/2 as they resist being squeezed into the same quantum state, which includes the same place in space and time. The Pauli exclusion principle largely governs the periodic table of the elements and is a dramatic consequence of the already stunning fact that electrons are described by the square roots of vectors (AKA spinors).

Now, with Einstein's new relationship between energy and momentum we are encountering yet another twentieth-century square root. Physicists, at first, saw no problem with simply ignoring negative energy states when they studied the energy of a particle like a photon or a meson. A meson is a particle which has no spin and a photon has spin-1, and for a meson or a photon the energy is always positive. The next step was to construct a theory of spin-1/2 (spinors) that fits together with Einstein's special theory of relativity. And so we come face to face with negative energy states, which brings us to one of the most revered figures in twentieth-century physics.

PAUL DIRAC

Paul Dirac was one of quantum theory's towering figures. For one thing, Dirac wrote *the* book on quantum physics, called *The Principle of Quantum Mechanics*, which became the standard reference.[8] This was the perfect authoritative treatment of the subject that had been built by the Bohr-Heisenberg school of thought. It gives the interrelationship between Schrödinger's wave function picture and the matrix formulation of Heisenberg. (We would recommend this book to anyone who would like to delve further, though it requires a good undergraduate training in physics).

Dirac's original contributions to quantum physics are among the greatest of the twentieth century. He notably considered the theoretical possibility of *magnetic charge*, "magnetic monopoles" that would act as pointlike sources for magnetic fields. Monopoles were absent in Maxwell's theory of electrodynamics, where magnetic fields are only generated by moving electric charges. Dirac discovered that the magnetic charge of a monopole and the electric charge of the electron are not independent but are inversely related via quantum theory. Dirac's theoretical work on monopoles conjoined quantum physics with the nascent field of topology in mathematics. Dirac's magnetic monopole has had a major impact on mathematics itself, and in many ways anticipates the style of thought and objects that are deployed later in string theory. Perhaps, however, one of the most profound discoveries of foundational physics in the twentieth century happened in Dirac's hands when he combined the electron (described by a spinor) together with Einstein's theory of special relativity.

In 1926, a young Paul Dirac sought a new equation for the spin-1/2 electron, one that went beyond Schrödinger's equation and that would be consistent with Einstein's theory of special relativity. To do this he needed spinors—the square roots of vectors—and he needed the electrons to have mass. But to make the equation work with relativity, he discovered that he needed to double the amount of spinor parts that an ordinary nonrelativistic electron has (i.e., each electron has two spinors).

Basically, a spinor is a pair of (complex) numbers, one representing the possibility of being spin-up, and the other of being spin-down. As usual, if we square these numbers, we get the probabilities of being spin-up or being spin-down. For this description of electrons to work with relativity, Dirac found he needed four complex numbers. The new equation, as you might have guessed, is called the "Dirac equation."

The problem is that the Dirac equation truly takes a square root in its full generality. The two original spinor parts that we started with, representing a spin-up electron or a spin-down electron, turn out to have positive energy, that is, we get the positive root of $E^2 = m^2c^4$, or $E = + mc^2$. But the two new spinor numbers, now required by relativity, take the negative square root and give us negative energy $E = - mc^2$. There was nothing Dirac could do about this; it was forced upon him by the symmetry requirements of relativity, essentially the requirements of the correct relativistic description of motion. This frustrated Dirac.

Indeed, this problem of negative energy is deeply buried in the fabric of special relativity and it simply cannot be ignored. Its thorniness became more severe as Dirac tried to construct a quantum theory of the electron. We can never avoid the negative sign of the square root (of energy) by simply dismissing it. Quantum theory with relativity apparently allows electrons to have both positive and negative energy values. We could say that the negative-energy electron is just another "allowed quantum state of the electron," but this would be a disaster as well. It would imply that ordinary atoms, even simple hydrogen atoms, all of ordinary matter, could not possibly be stable. The positive-energy electron, of energy mc^2, could then emit photons, adding up to an energy of $2mc^2$, and become a negative-energy electron, of energy $-mc^2$, and begin its descent into an abyss of infinite negative energy (as the particle's momentum would increase, the magnitude of the negative energy would become larger and larger).[9] Evidently, the whole universe could not be stable if the negative energy states truly existed. The new requisite negative-energy electron states were now a prime headache.

Dirac, however, soon had a brilliant idea that solved the problem of the negative energy abyss. As we have seen, Pauli's exclusion principle says that *no two electrons can be put into exactly the same quantum state of motion at the same time*. That is, once an electron occupies a given state of motion and spin—a quantum state of motion, like an orbital in an atom—that *state is filled*. No more electrons can join in (of course, we recall that the quantum state of motion can have two electrons, one with spin-up, and the other with spin-down). Dirac's idea was that *the vacuum itself is completely filled with electrons, occupying all the negative energy states*. All the negative energy levels in the whole universe are therefore filled, each with a single spin-up and spin-down electron. Then positive-energy electrons, such as in atoms, could not emit photons and drop down into these states since they would be *excluded* from doing so by the Pauli exclusion principle. In effect, the vacuum in this picture would become one gigantic inert atom, like a humongous argon or radon atom, with all the possible states of negative energy, for any momentum, already filled.

Dirac's idea of a vacuum in which all negative energy levels are already occupied by electrons seemed to put an end to the issue of the negative-energy catastrophe once and for all. It is a bizarre notion that the vacuum is filled up with negative-energy electrons, but it seemed to stabilize the world against falling into the negative energy abyss.

We call this view of the vacuum the "Dirac sea." The Dirac sea is not empty but rather is a completely filled "ocean" that metaphorically represents the infinity of filled negative energy levels (see figure 32). At the initial conception of the idea, Dirac thought that this was the whole story—until . . .

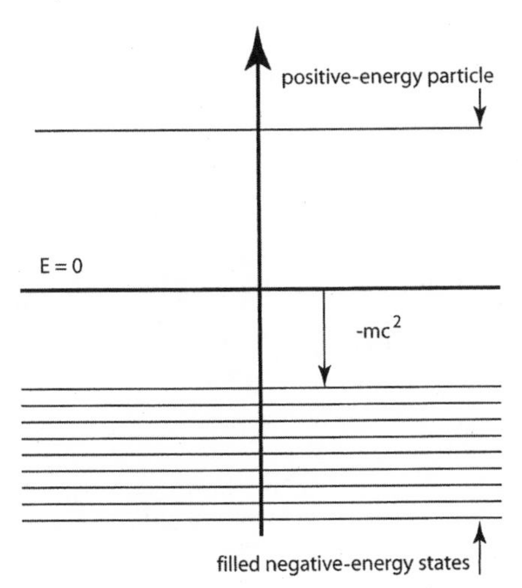

FIGURE 32: The Dirac sea. All the allowed negative energy levels that are predicted when relativity is combined with quantum theory are filled. The vacuum is like an enormous inert element, like neon. This implies that positive-energy electrons are stable and won't tumble down into empty negative energy levels.

FISHING THE DIRAC SEA

Dirac soon realized that the story didn't end there. He discovered that it was theoretically possible to "excite" the vacuum. This means that physicists could arrange a collision in which they pull a negative-energy electron out of the vacuum, much like a fisherman pulls a deep-sea fish into his boat. Now, usually when a high-energy gamma ray collides with a negative-energy electron in the vacuum, nothing happens. A single gamma ray hitting a negative-energy electron cannot raise it out of the vacuum because such a process wouldn't conserve all the necessary quantities that physics demands be conserved, in other words, momentum, energy, angular momentum. However, if there are other particles also participating in the collision (like a nearby heavy atomic nucleus, to recoil slightly and conserve the overall momentum, energy, and angular momentum of the participants in the collision; we call this a 3-body collision), then the electron could be ejected out of the Dirac sea into a state of positive energy. The gamma ray could then successfully eject an electron out of its negative energy state, and into one of positive energy, that could register in the physicist's instruments.

However, Dirac realized that this collision would leave behind a *hole in the vacuum*. The hole, however, would represent the *absence of a negative-energy electron*. This means that the hole *actually would have a positive energy*. However, the hole would also represent the *absence of a negative electrically charged electron*, and hence the hole would be a *positively charged particle* (see figure 33).

Dirac had predicted the existence of something totally new and totally bizarre: *Antimatter.* An antiparticle is the "hole" in the vacuum representing the absence of a negative energy particle (hence having positive energy). Every particle in nature has a corresponding antiparticle. We call the antiparticle of the electron the *positron.* The positron is a positively charged particle with positive energy and is otherwise indistinguishable from the electron, though it is simply a hole in the vacuum of filled negative energy levels The laws of special relativity require that the hole in the vacuum, at rest, must have an energy of exactly $E = +mc^2$, where m is *exactly the electron* mass. Positrons were predicted by Dirac and they must exist if both quantum theory and special relativity are true.

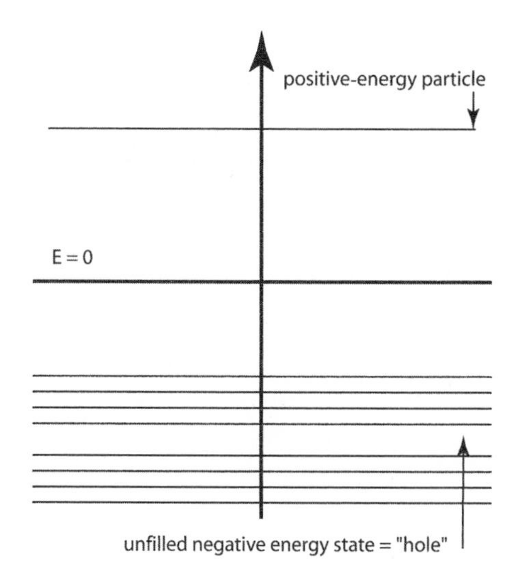

FIGURE 33: Dirac's sea leads to the prediction that a negative-energy electron can be ejected out of the vacuum by the collision of a photon with a nearby atom. The hole left in the vacuum is the absence of a negative-energy, negatively charged electron, and therefore appears as a positive-energy, positively charged particle with identical mass to the electron. Dirac thus predicted the positron, and the phenomenon of electron-positron pair creation. The positron was discovered experimentally a few years later by Carl Anderson.

In fact, Dirac was quite bothered by the positron. The erudite culture among theoretical physicists of that time was one of *minimalism*—trying to explain what is and not invoking extra baggage. Dirac didn't like his new positron at first and held out hope that he could somehow bend or twist it into an explanation of the already known proton, the much heavier positively charged particle that forms the nucleus of the hydrogen atom. Alas, the proton is two thousand times heavier than the electron, and the symmetry of relativity insists that the positron—the hole left behind in the vacuum by the missing negative-energy electron—must definitely have the same mass as the electron.

Positrons were subsequently discovered in an experiment in 1933 by Carl Anderson. They are produced by "cosmic rays"—energetic particles coming from space—that collide with the negative-energy electrons in the vacuum in the presence of heavy atoms to produce a positron and an electron.[10] The produced positron and electron were observed in a *cloud chamber*. A cloud chamber is an early kind of particle detector that contains a gas, for example, nitrogen or argon, though even air supersaturated with a water or alcohol vapor works. As an electrically charged particle travels through the chamber, it leaves behind a trail of tiny vapor particles that are induced to grow to visible sizes and can then be photographed. Typically, the original cosmic ray particle passes through a thin plate of material and, in doing so, knocks out a negatively charged electron together with its positively charged hole (the positron). A strong magnetic field is applied to the cloud chamber, causing the particle motion to curve in a way that reveals its electric charge. Anderson observed pairs of electron and positron as two separate curling tracks, curling in the opposite sense, in the cloud chamber, several years after Dirac's theory had predicted them. The positron mass could be measured and was indeed the same as that of the electron, as special relativity requires.

Such events confirm the existence of antimatter, for soon after the existence of the positron (antielectron) was confirmed, the negatively charges antimatter twin of the positive proton was seen. Today, all the known particles: quarks, charged leptons, neutrinos, and so on, are confirmed to have antimatter twins.

The discovery of antimatter is one of the most stunning theoretical and experimental achievements in human history. Antimatter will "annihilate" matter when the two collide, as the positive-energy electron jumps back into the hole in the vacuum, usually emitting gamma rays to conserve energy and momentum. The annihilation produces a lot of energy (at rest, electron positron annihilation would release $E = 2mc^2$ by direct conversion of all the rest-mass energy of the two particles into gamma rays). In annihilation of matter with antimatter, the particles simply jump back down into their holes in the Dirac sea, and the energy is transformed into other low-mass particles.

In the very early universe at ferociously high temperatures, there was exact equality in the abundance of particles and antiparticles. If this perfect symmetry had persisted, all the matter and antimatter would have annihilated into photons, and we would not exist. For reasons that remain mysterious today, there is no antimatter left in the universe, but we do exist, that is, there is matter today but no antimatter. In the very early universe a tiny asymmetry between the abundance of matter and antimatter somehow developed. As the universe cooled, most of the matter annihilated with antimatter, leaving the small excess of matter that, today, constitutes all the visible matter in the universe (including us). The precise mechanisms by which this asymmetry between matter and antimatter happened is as yet unknown and must lie in new physics that has not yet been discovered.[11]

Positrons, and other antiparticles, can be artificially produced by particle accelerators. Antimatter is a useful commodity and is already "paying rent." Positrons are naturally generated from radioactive disintegration of certain atomic nuclei and have found a use in positron emission tomography (PET) scanners, a form of medical imaging. It is estimated that the cash flow generated by this one activity, a by-product of pure and basic research, is much larger than the cost of funding all the science of particle physics today. It is unclear if the future utility of synthesized antimatter will expand to warp-drive starship engines, but eventually it will likely find some more practical applications. One "application," shown in a ridiculous and scientifically inaccurate movie, was a plot in which antimatter was stolen from CERN in order to blow up the Vatican. While we don't know what the ultimate good and practical applications of antimatter will be we're sure that one day the government will tax it.

Corresponding to *every* particle there is an antiparticle in nature. Corresponding to protons we have antiprotons, to neutrons we have antineutrons, to top quarks we have antitop quarks. When we make top quarks at the Fermilab Tevatron or the CERN LHC, we make them in pairs—top plus anti-top. We literally go fishing and pull the negative-energy top quark out of the deep depths of the vacuum. This leaves behind a top quark hole (the anti-top), and we see the pair, quark and antiquark, produced in our detectors.

Particle physicists are fisherman on the great Dirac sea. They now seek a whole new species of fish from the depths of the Dirac sea. To do this they have built a massive fishing pole, the Large Hadron Collider (LHC) in Geneva, Switzerland. What will they find?

> ALTHOUGH you hide in the ebb and flow
> Of the pale tide when the moon has set,
> The people of coming days will know
> About the casting out of my net,
> And how you have leaped times out of mind
> Over the little silver cords,
> And think that you were hard and unkind,
> And blame you with many bitter words.
>
> William Butler Yeats, "The Fish" (1898)[12]

THE TROUBLE WITH THE ENERGY OF THE DIRAC SEA

New ideas usually arise from old problems. Dirac invented his sea to solve the problem of the negative-energy abyss problem and was led to antimatter. But now, the Dirac sea itself raises a central problem in physics. It has to with gravity, but it also has to do with the quantum theory, so we'll pause to think momentarily about it now.

Gravity is that ubiquitous force that is produced by anything that has mass, energy, and momentum. These three attributes are certainly present in each of the (negative-energy) electrons that fill the great Dirac sea. In fact, we apparently have an infinite amount of negative energy in the Dirac sea: If we start adding up all the negative energies of each of the particles in the sea, we rapidly get an uncontrollably large negative sum. If we stop doing our addition at an arbitrary energy level whose energy value is $-\Lambda$ ("lambda"), then we find that the total energy per unit volume, or "vacuum energy density," is about $\rho = -\Lambda^4/\hbar^3c^3$(this is the amount of vacuum energy of the Dirac sea per unit of volume). This turns out to be a very large (negative) number. For example, if we chose an energy scale equivalent to the proton mass, for the value Λ, this would be an energy density that is a million, trillion times greater than that of ordinary water. Such an energy density is not found anywhere in today's universe.

This is a mathematical "runaway" of our arithmetic sum. It is like adding together all of the negative integers. For example, try adding:

$$-1 - 2 - 3 - 4 - 5 - 6 - 7 - 8 - 9 - 10 - 11 = -66$$

We stopped at -11 (the analogue of $-\Lambda$) and we got the result of -66. Here we are simply imitating the calculation of the vacuum energy density from 11 filled negative-energy quantum electron states in the Dirac sea.[13]

So, to state what we have done, if we "add up the first 11 negative integers, we get a result of –66." If we keep on going, let's say up to 100 negative integers, we get the result –5050. If we add up 1000 negative integers we would get –500500. The value of the energy would grow more and more negative the larger we make the cutoff. Mathematicians have a name for this: they say we are adding up a "series" of numbers, and since the sum grows bigger and bigger, they say the series is "divergent."

When we calculate certain things in quantum theories, such as certain properties of electrons and photons, we get divergent results. Even though almost all the things we compute get sensible answers—in fact, answers that are consistent with experiment—a few simply yield mathematical nonsense. As we have just seen, when we are calculating the vacuum energy density in quantum theory, we get a divergent series. If we didn't halt the sum at some cutoff point, our result would be minus infinity for the vacuum energy density, and that's just plain nonsense. If the vacuum had such an energy density, our universe would crumple up into an infinitesimal pinpoint, crushing everything into nothingness, including you, me, and all our literature. Getting this nonsensical answer of a negative, infinite vacuum energy density tells us that something fundamental is wrong, or at least missing, in our theory.

Nonetheless, our theory is still pretty good—it gives us the correct and dramatic prediction that antimatter does exist! And the whole structure of "quantum electrodynamics"—the theory of the quantum electron interacting with the quantum photon—gives exact and precise predictions for almost all the processes involving electrons and photons. For example, because the electron is electrically charged, and it spins, it is therefore a little magnet. We can successfully compute and measure the magnetic field around an electron to a precision of one part in a trillion. The agreement between theory and experiment is stunning. These annoying infinities occur only in a few places, but otherwise our theory is predictive and spectacularly successful, so we want to keep it. The problem is to understand what these infinities are telling us and how to fix them.

As far as the negative infinite vacuum energy density in the Dirac sea is concerned, we still, to this day, have a conundrum. The problem is that the vacuum energy density affects the whole universe through gravity. The expansion and "size" of the universe are controlled by all its matter contents through energy, mass, and momentum. The contents of the universe include the vacuum. From the observations of the rate of expansion of the universe it is inferred that there may indeed be a tiny vacuum energy density, but it is evidently positive and not negative. We call this the "cosmological constant," and it is extremely small. In our formula it corresponds to a value of the cutoff, Λ, of about 0.01 electron volts, a very small cutoff energy.

But how do we calculate Λ? Our quantum theory doesn't tell us what Λ is, much less how to proceed. So if we chose Λ to be about the mass of the electron (just a guess), which is certainly a scale at which quantum electrodynamics is a valid description of nature, we would predict a Λ of about a million electron volts. This would imply a vacuum energy density that is negative and 10^{32} times bigger than what is observed. This would leave us with a monstrous discrepancy between theory and observations.

In fact, this problem has nothing to do with what the precise value of the "cosmological constant" actually is. All we need know is that the cosmological constant is infinitesimally smaller than what we "computed." Usually physicists argue that quantum electrodynamics (or something like the Standard Model, which contains quantum electrodynamics and all other forces and particles, quarks, gluons, neutrinos, other photon-like objects, etc.) should be valid up to the scale of quantum gravity, called the "Planck scale." This is the mass scale we get by combining Newton's gravitational constant, Planck's constant, and the speed of light. It is believed to be the energy scale at which quantum gravity is in effect, and space and time will become a kind of quantum foam, or quantum spaghetti, perhaps made of strings. The Planck scale corresponds to about $\Lambda \approx 10^{19}$ giga electron volts. Using the Planck scale as a cutoff, we predict a cosmological constant that is 10^{120} times bigger than what is observed. This is often called the biggest mistake in all of physics! This enormous mismatch between our predicted vacuum energy density and the observed cosmological constant means that something is very much out of whack with our quantum theory of the electron and photon when combined with gravity. What could it be?

One thing we have left out from our calculation of the vacuum energy is the effect of the photon. The photon is a boson, that is, it has integer spin and is not subject to Pauli's exclusion principle (see appendix on spin). We may therefore think that bosons do not reside in Dirac's sea, but it turns out that bosons also have antiparticles. But the main distinction between bosons (integer spin) and fermions (half-integer spins) is really more subtle: bosons *do not have negative energy states*.[14] The energy of a boson is always positive. Moreover, the vacuum according to a quantum theory of bosons also has a divergent vacuum energy density, *but* it turns out to be positive. This happens because, essentially, quantum theory forbids the bosonic particle from standing exactly still. It must always "twitch" (or, perhaps, "twitter," in a modern parlance). Therefore, even in its ground state (i.e., the vacuum), a boson has a nonzero positive energy.

So photons have positive vacuum energy and electrons have negative vacuum energy. When we compute the vacuum energy of electrons and photons, we must add together the negative Dirac sea energies from the electrons and the positive energies from the photons. It turns out we still get a net negative answer, and the answer is still infinite. So we still have a big problem. Maybe with more photonlike fish in the sea we can cancel the negative energy of the electrons?

SUPERSYMMETRY

The vacuum energy calculation only gets worse when we include the muon, the neutrinos, then the tau lepton, the quarks and gluons, then the W and Z bosons—even the undiscovered Higgs boson. These are all of the creatures that inhabit Mother Nature's known zoo of particles. Each of these contributes its own vacuum energy, negative for the fermions and positive for the bosons, with an uncontrolled infinite result. What is lacking here is not a better calculation but rather a new physical principle that tells us how to compute the vacuum energy density of the universe—and to this day we don't have one.

There is, however, a remarkable symmetry that one can create for a "toy" quantum theory that actually does allow us to compute the cosmological constant and actually gives us a sensible mathematical result: zero! We can actually make a direct connection between the fermions in a theory with the bosons, and vice versa. We do this by introducing an imaginary extra dimension that only Lewis Carroll could have imagined before the modern era. And this new dimension itself behaves like a fermion—it "excludes" à la Pauli, any more than one single step into it.

Once we take one step into the new dimension, then we are done (much like putting a single electron into a state and we cannot put a second one into the same state). But when a boson takes this one step, it turns into a fermion; and when a fermion takes such a step, it becomes a boson. So, for quantum electrodynamics, if such a weird dimension existed and if we took one step into it, like Alice in *Through the Looking Glass*, the electron would turn into a kind of bosonic particle called a *selectron*. The photon, upon taking such a step, would turn into a fermionic particle called a *photino*.

This weird new dimension of space represents a new kind of physical symmetry that we are conjuring up mathematically, and it is called *supersymmetry*.[15] In supersymmetry, every fermion particle has an associated boson partner particle, and vice versa. The whole particle content of the theory is doubled. The relationship between a particle and its superpartner is akin to a particle and its antiparticle. The grand effect of this is, as you may have guessed, that when we compute the vacuum energy density, we now get the positive vacuum energies of the bosons exactly cancelling against the negative Dirac sea energies of the fermions. Voila! The cosmological constant is then identically zero.

So can supersymmetry solve the real-world problem of the vacuum energy? Perhaps, but it isn't clear how. The problem is twofold: first, we do not observe a bosonic superpartner of the electron.[16] Moreover, there is evidence of a small positive cosmological constant in nature, and exact supersymmetry would not yield a nonzero value. However, any symmetry (such as the symmetry of a perfect spherical orb of clay) can be "broken" (by squashing the orb of clay with your fist). Physicists have a deep-seated love of symmetry: powerful mathematical symmetries have always become the ingredients of our best-loved and cherished theories. Therefore, most physicists are hoping that there does indeed exist supersymmetry in nature but that some dynamical mechanism (like a fist onto a ball of clay) has "broken" the supersymmetry, and that we will only see it as we go to very high energies with new particle accelerators, like the LHC. The breaking of supersymmetry implies that the superpartners of the electron and the photon, the selectron and the photino, are very heavy particles and won't be seen until experiments of sufficiently high enough energy, Λ_{SUSY} (SUSY is the standard abbreviation for supersymmetry), can manufacture them.

Alas, the act of breaking supersymmetry causes the vacuum energy problem to return. Now we find that the vacuum energy is determined by the scale of supersymmetry breaking, as $\Lambda^4_{SUSY}/\hbar^3c^3$ If the scale is near Fermilab Tevatron or CERN Large Hadron Collider energy scale—one to ten trillion electron volts— then we still have a vacuum energy, or cosmological constant, that is 10^{56} times too large. That's a significant improvement over 10^{120}, but its still a problem. So, SUSY, at least in its most direct application, doesn't help solve the vacuum energy crisis. What does?

HOLOGRAPHY

Is something going haywire in how we are counting the fish in the Dirac sea? Perhaps we are overcounting them? Ultimately, our counting is allowing for extremely tiny fish, that is, extremely short-wavelength negative-energy electrons. This gets to be a very tiny sub-sub-nuclear energy scale when we make the unknown cutoff energy large. Perhaps such tiny states are not really there after all?

A radical new idea has emerged over the past decade that scientists have overcounted the number of fish in the Dirac sea because the fish aren't filling a three-dimensional sea of space—rather, the world is a hologram! A hologram is a projection of all of space onto a space of smaller dimension, as if projecting all of 3D space onto a 2D sheet of paper. The rule is that whatever is happening in the 3D space can be completely described by what is happening in the 2D world on the surface of paper. So 3D space is not filled with fish in the way we have counted it. In short, all of those negative energy levels are just an illusion. In the hologram theory, space is very sparsely filled with fish. In fact, the fish themselves are 2D objects. The result is that we get a vacuum energy that is significantly reduced and may even explain the observed tiny cosmological constant. We say "may explain," because this is a work in progress. There is no exact holographic theory yet.

This new holographic idea comes from certain discoveries in string theory whereby a definite holographic connection can be made (the most precise and original of these is called the Maldecena, or AdS/CFT conjecture).[17] We will return to this new holographic idea, whether fanciful or profound, in the next chapter, but the general sense of it may well be a kind of "dream logic."

FEYNMAN'S SUM OVER PATHS

Some particles are their own antiparticles. We call these "self-conjugate particles." For example, the photon is self-conjugate. The π^+ meson has for its antiparticle the π^- meson, but the π^0 meson is self-conjugate and is its own antiparticle. But hold on, didn't we say that mesons always have positive energy? Then why do mesons have antiparticles, if they are not holes in the Dirac sea?

Antimatter is a general phenomenon that applies to both bosons and fermions. While Dirac's sea is very tangible and serves us well in understanding fermions, it was Richard Feynman who arrived at another, and perhaps more general, way of looking at it. Feynman's idea helps us resolve many of the unsettling puzzles we encounter throughout quantum physics, such as EPR paradoxes, among other things. Feynman, building on some ideas of Dirac, in his doctoral dissertation at Princeton, reformulated the quantum theory in a new and stunningly useful way. To appreciate Feynman's innovation, let's first recall Newton's notion of particles and then Schrödinger's wave function.

Newton said a particle should be described by stating where it is in space, x, at time t. This is a trajectory, or a mathematical function called $x(t)$. The actual trajectory that the particle takes is then determined by solving Newton's equation of motion. But Schrödinger and his colleagues formulated quantum physics in a completely different way: a particle does not take a definite path; rather, a particle is described by a wave function $\Psi(x, t)$ that gives the "quantum amplitude" to find the particle at x at a given time t. The square of the amplitude is the probability that the particle will be at position x at time t.

Then enters Feynman. Feynman said that Schrödinger is ultimately right, but let's descend to a more fundamental level and ask how a particle, released from some initial place x_0 at some initial time t_0 would end up later in a Schrödinger's wave function $\Psi(x, t)$. Feynman gave us the answer: The wave function is just the sum of all the possible paths the particle could take in getting to x at time t. So what exactly are we summing? Each path involves a mathematical factor called the "phase." The "phase" is a function of any given path that starts at, say, x_0 at time t_0 and ends up at x at time t. Feynman tells us the rules for computing this phase. There are usually an infinite number of possible paths, and we have to add up the phase factors for each of them, but we have sophisticated mathematical methods for doing this. While this is seemingly daunting, this approach actually leads to something very manageable in many cases, and it give us a much clearer picture of how things are happening throughout space-time in quantum physics.[18]

In fact, Feynman's sum over paths comes directly out of Young's experiment. In the famous double-slit experiment, there are only two paths to add up

(1) An electron is released from the source and travels through slit 1 and then to point *x* on the catcher screen (for which we get a "phase" F_1; Feynman tells us how to compute this phase factor).
(2) An electron is released from the source and travels through slit 2 and then to point *x* on the catcher screen (for which we get a "phase" F_2).

So Feynman tells us that the amplitude to find the electron on the catcher screen at any point is just $F_1 + F_2$. This is Schrödinger's wave function. The probability is just the square of this quantity, $(F_1+F_2)^2$. And if we plot the resulting probability distribution, we get the now familiar interference pattern (see figure 17, on page 106), in perfect agreement with the experiment. It arises because nature explores all possible paths for the motion of a particle through space and time (in this case, two of them) and adds up the amplitudes for all such paths. The amplitudes interfere when we square to get the probability.

Feynman's path integral shows us immediately that if we cover one of the slits, say, slit 2, there is now only one path the electron can take and the amplitude is now F_1. In this case the interference pattern completely disappears (see figure 18, on page 107).

The Feynman sum over paths, also called the "path integral," clarifies (at least to some minds) the EPR experiment and the entanglement probed by Bell's experiment as well; for EPR, when a radioactive particle decays into a pair of particles, one spin-up and the other spin-down, there are then two "paths" throughout space-time to consider. One path (call it "A") delivers the spin-up particle to remote detector 1 and the spin-down particle to remote detector 2; the other path (call it "B") delivers spin-down to detector 1 and spin-up to detector 2. Each path has a certain "quantum phase" or "quantum amplitude." To get the total amplitude, we add up the two path amplitudes A + B. This indeed leads to the "entangled states" at time of detection. But what is new is now a matter of interpretation: when we make a measurement at detector 1, we are simply discovering which of the two paths throughout space and time the system chose to take. That is, if we measure spin-up at detector 1, then the particles took the first path "A." If we measure spin-down at detector 1, then the universe took path "B." There seems to be no disturbing instantaneous change in the wave function throughout space anymore, because the particular path contains the information (the correlation) as to what will happen in detector 2 if a given result is observed in detector 1. This is no more spooky than the classical blue or red billiard ball mailed to us on Earth and to a friend on Rigel 3.

Indeed, quantum physics is awesome and gives us goose-bumps, since nature mysteriously probes all possible paths and gives only the aggregate sum, with interference. But the space-time description of paths seems to have removed the creepy idea of things traveling faster than light that so bothered EPR. So how from the point of view of the sum over paths can we understand antimatter?

Feynman interpreted the positive-energy particles as moving along paths that go forward in time. On the other hand, the negative-energy particles he interpreted as moving *move backward in time.*

The production of an electron and positron is shown in figure 34. Here we see a photon producing the electron-positron pair at a point (A) in space-time. But from the path integral point of view we see the positive-energy electron going forward in time, but the antiparticle has arrived at point (A) by coming in from the future! Then in the distant future at a point in space-time (B), the electron collides with a positron and annihilates back into a photon. But from the path integral point of view, the positive-energy electron moving forward in time has turned around and become a negative-energy particle (the positron) traveling backward in time at point (B)!

Feynman was bemused by this and reportedly phoned his Princeton thesis advisor, the venerable John Archibald Wheeler, late at night to announce that there is only one electron in the whole universe! According to Feynman, the sole electron propagates forward in time where it emits a photon and turns, coming back into the past as a negative-energy particle (an antiparticle). This would appear to an alien being at the end of the universe like an annihilation event of an electron with a positron. Then the negative-energy particle propagates back to the origin of the universe where it collides with a photon (which would appear to an alien being like an electron positron pair creation event) and again turns around coming back to the future as a positive-energy electron, and so on and so on.

There is a deeper reason for this seeming madness of thinking of antimatter as matter traveling backward in time. It turns out that it is precisely the balance between quantum paths of particles moving forward in time and negative energy paths moving backward that makes physical signals "causal." That is, this prevents signals from traveling faster than the speed of light. The whole fabric of space and time—causality and relativity—is therefore interwoven into the existence of antimatter in quantum physics. And it all must involve bosons as well as fermions. If particles were found to have features different than their antiparticle counterparts, such as mass or spin or (magnitude) of electric charge, then the path integral predicts that signals could in principle be transmitted at speeds faster than light. We see no such evidence of such particle-antiparticle differences.

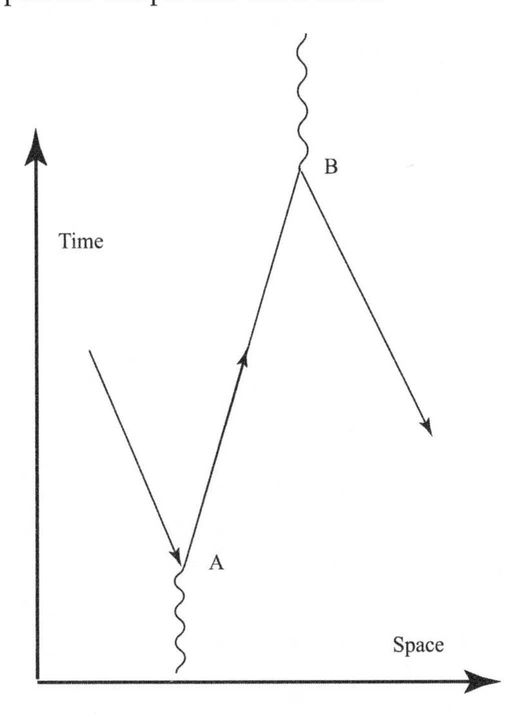

FIGURE 34: Production and annihilation of antimatter in Feynman's view. A photon collides at event (A) with a negative-energy electron coming in from the future (a positive-energy positron), which turns around and heads into the future as an electron. This appears to us as the photon producing a pair of electron and positron. In the distant future the electron emits a photon and turns around heading back into the past as a negative-energy electron (a positive-energy positron). We observe this as the annihilation of an electron (matter) with a positron (antimatter).

So obviously we wonder, can these particles approaching us from the future allow us to see into and predict the future? Physics says no, because of the strict enforcement of causality by the mere existence of antimatter. Signals cannot be transmitted faster than the speed of light because the sum of all paths for such a signal adds up to zero, and this is a consequence of the existence of antimatter with its precisely opposite properties between particle and antiparticle.

CONDENSED MATTER PHYSICS

The quantum theory has profound and extremely useful applications to the world's materials. In fact, it largely allowed us to understand for the first time what the states of matter are and how they work, as well as the phases of matter and the subtle magnetic and electrical conduction properties of matter. As in the case of the periodic table of the elements, quantum physics has paid off handsomely, while both explaining the structure of everyday matter and enabling the creation of new technologies. It has spawned the new area of "quantum electronics" and has revolutionized our everyday lives in ways that were unimaginable a century ago. Let's focus on one major subdiscipline of this vast topic, which deals with electrical currents flowing in materials.

THE CONDUCTION BAND

When atoms form a solid, they become pushed together into close proximity to one another. The electron wave functions of the highest occupied orbitals of the atoms begin to blend together (while deeper filled orbitals are essentially unaffected by the formation of the solid material). Electrons in the highest orbitals begin to jump from one atom to another. In fact, the highest orbitals lose their identity and are no longer localized around a given atom, as the electrons begin to stray throughout the whole material. The highest orbital states of the atoms blend into a collection of extended states of motion for the electrons, called the *valence band.*

Suppose we have a crystalline material. Crystals can come in many different forms, and each crystal is defined by its crystal lattice. The possible lattices and their properties have been classified by physicists.[19] The electrons that begin to stray throughout the crystal have wave functions with very long wavelengths in the valence band. The straying electrons fill these states of motion according to the Pauli principle: at most two electrons, one spin-up and the other spin-down, can occupy an allowed quantum state of motion. The very long-wavelength states are much like electrons moving in free space, and there is then no interference from the crystal lattice. These states have the lowest energies and fill up first. The straying electrons fill more and more states until their quantum wavelengths become comparable to the distance between the atoms.

However, the electrons scatter by electromagnetic interactions off the atoms in the lattice. Crystal lattices act like an enormous Young's interferometer, but with many, many slits—the slits are the scattering centers of the lattice, so there is one slit for each atom. Thus, the electron motion involves an enormous amount of quantum interference.[20] The interference occurs when the electron quantum wave length is comparable to the spacing between the atoms. States that would have electron wavelengths close to these special values of wavelength (or momentum) interfere destructively and are therefore blocked out.

This interference causes the formation of a *band structure* of the energy levels of the electrons in a solid. An energy gap, called a *band gap*, forms between the lowest band (the "valence band" full of the lowest-energy, straying electrons) and the next highest energy band. The electrical conductivity of the material critically depends on the band structure. This leads to three distinct possible ways in which the material can behave in its conduction of electricity.

1. Insulators: If the valence band of a material is completely filled up with electrons, and we have a big energy gap below an unfilled higher energy conduction band, then we have an *electrical insulator*. Such a material will not conduct electricity (like glass or plastic). This happens typically with materials that have nearly filled atomic shells, such as halogens and ionic molecules or the noble gases. Under this circumstance, electrical current cannot flow because there is no room for the electrons in the valence band to "slosh around." That is, in order to move, the electrons would have to jump through the large energy gap into the conduction band and that would require too large an amount of energy.[21]

2. Conductors: If the valence band is partially filled, then electrons can easily move into new states of motion. This makes a good conductor of electrical current, that is, electrons can easily slosh around and conduct electrical current. This happens when, typically, there are many electrons available to stray away from their atomic orbitals. Thus, those atoms with unfilled highest orbitals that typically donate electrons to form chemical bonds, such as in the alkali and the heavier metallic atoms, are good electrical conductors. Moreover, it is the scattering of light off these loose conduction band electrons that makes a metal conductor shiny. As the conduction band begins to fill up, the material becomes a poorer conductor of electricity, approaching the conditions of an insulator.

3. Semiconductors: If the valence band is nearly completely filled, or if the conduction band has relatively few electrons, then the material cannot conduct much electrical current. However, if the energy gap isn't too large, about three electron volts or less, we can coax electrons more easily up into the conduction band. When this kind of energy band structure is present, we have a semiconductor. The marginal ability of semiconductors to conduct electrical current makes them remarkable: we can strongly influence the conductivity in different ways, allowing ourselves to make "electronic switches."

Semiconducting materials are typically crystalline solids, such as silicon (sand). The conductivity—the ability to conduct electrical current—of semiconductors can be drastically changed by adding other elements, "impurities," which is called "doping." The semiconductors with a few electrons in a conduction band are called "n" type materials. They are typically engineered by adding to the material atoms that lend more electrons to the valence band and are thus populating the conduction band with electrons. Semiconductors with nearly filled conduction bands are called "p" type and can be engineered by adding doping atoms that draw electrons out of the valence band.

In a p-type material, we have a virtual absence of electrons filling the valence band to complete an insulator. This absence of electrons is like the "holes" we encountered in the Dirac sea, the positrons, so the holes in a semiconductor thus act like positively charged particles that can carry electrical current. A p-type material is, therefore, a little Dirac sea, created in the laboratory. Holes, however, actually involve the motion of many electrons, so they act as though they are much heavier than a single electron and are less efficient current carriers than electrons themselves.

DIODES AND TRANSISTORS

The simplest example of a gadget we can make with semiconductors is the diode. A diode conducts electricity well in one direction but acts like an insulator in the other direction. We generally get a diode when we place a p-type material in contact with an n-type material, forming a "pn junction." It is easy to get the n-type electrons in the conduction band to jump across the junction down into the valence band of the p-type material. This is like particle-antiparticle annihilation in the Dirac sea, but notice that the electrical current is only flowing in one direction.

If we try to reverse the flow of current, we find it is difficult, as we tend to suck the conduction band electrons out of the junction, and no electrons are available to replace them from the p-type material. So in a diode, provided you don't apply too much voltage (you can easily burn up a semiconductor by forcing a large current through it), the current can easily flow only in one direction. Diodes play a major role in the design of electronic circuits for many electronic devices.

In 1947, John Bardeen and William Brattain, working in a group led by William Shockley at Bell Labs, made the first transistor, called a "point-contact transistor." This is a generalization of a diode to a triple junction of semiconducting materials. It allows one to control the current flow through the device by varying the voltage between the input and a center material called the "base." Varying the voltage between the first two materials affects the conduction band so as to allow or disallow current flow between the first (emitter) and third (collector) material. The transistor is probably the most important device humans ever invented and it won the Nobel Prize for Bardeen, Brattain, and Shockley in 1956.[22]

PROFITABLE APPLICATIONS

So what good is this? Schrödinger's powerful equation, yielding the rules by which the wave function can be calculated, began as the brainchild of pure reason, and few could have imagined that it could run expensive machinery or fuel national economies. But when applied to metals, insulators, and (most profitably!) semiconductors, this equation enabled physicists to invent switches and control elements that evolved to a million transistors on a chip. These made powerful computers and devices that control huge instruments, such as particle accelerators, automobile assembly plants, video games, and airplanes landing in foul weather.

Another fair-haired child of the quantum revolution is the ubiquitous laser, used for supermarket registers, eye surgery, metal cutting, and surveying, and also as a tool to learn more about the structure of atoms and molecules. The laser is a "flashlight" that emits photons of light—all of the exact same wavelength.

We could go on for pages about the technological miracles that owe their existence to the insights of Schrödinger, Heisenberg, Pauli, and others, but let's mention just a few. The first is the scanning, tunneling microscope, which gave us a magnification thousands of times more powerful than the best electron microscope (a quantum invention itself, by the way, based on the wavelike features quantum theory ascribed to electrons).

The tunneling principle is quintessential quantum theory. Think of a small smooth bowl sitting on a table, within which a polished steel ball rolls up and down. Classically and in the theoretical absence of frictional forces, the ball is trapped in the bowl for all eternity, rolling down one side and up to the same height on the other side. Perfectly Newtonian. The quantum version of this phenomenon would be an electron trapped in a box with gridded walls carrying a repelling voltage that requires more energy to scale than the electron has. So the electron approaches the grid, is repelled, collides with the opposite wall, is repelled again, and rolls back and forth . . . endlessly? No! Sooner or later, in the spooky quantum world, the electron appears outside the box.

Can you imagine how unsettling this is? From the classical point of view, you'd say that it magically tunneled through the wall, quite as if our metal ball had escaped, Houdini-like, from the confines of its bowl and bounced on the coffee table. Schrödinger's equation endows the problem with wavelike probability aspects so that, at each collision of the electron with the wall, there is a small probability that the electron will penetrate the wall. Where does it get the energy? Not a good question, as the equation does not describe a trajectory through the impossible wall—only a probability that the particle is inside and a probability that it is outside. Disquieting as it may seem to the Newtonian mind, tunneling works. In fact, by the 1940s it became a cottage industry, explaining some heretofore inexplicable phenomena of nuclear physics. Pieces of the nucleus actually tunnel through the barrier that holds them together and breaks the nucleus into smaller nuclei. This is *fission*, the basis of nuclear reactors.

Another practical device using this weird effect is an electronic switch called a Josephson junction, after its brilliant, quirky inventor Brian Josephson. The Josephson junction operates at temperatures close to absolute zero, where quantum superconductivity adds to its exotic character. Someone called it a superfast, supercold, superconducting quantum tunneling digital electronic device! That sounds like something out of a Kurt Vonnegut novel, but it exists nonetheless, and it is able to switch an electronic current at a speed of many trillion times a second. In the exploding era of high-speed computers, switching speed is the name of the game. Why? Calculations involve bits, a bit being either a zero or a one. Algorithms convert a series of zeros and ones into numbers that can add, subtract, multiply, divide, integrate, differentiate, scratch your back, and cut your toenails. So switching from on (one) to off (zero) is the primordial act. And the Josephson junction switch does it best.

Quantum tunneling has led to other scientific breakthroughs. The application of quantum tunneling to microscopes enabled humankind to "see" individual atoms, as, for example, in the sweeping organization of a double helix of intertwined atoms that compose the DNA code—the repository of all information that defines any living organism. The scanning tunneling microscope (STM), invented in 1980, does not observe its objects by the light of a microscope lamp (as in the familiar light microscope) or by the probability waves of a beam of electrons (as in the electron microscope). Its governing principle is a super-sharp needle probe that moves over the contours of the object being observed, carefully calibrated to hover about a hundredth of a millionth of an inch above the surface. The gap is small enough so that an electric current (from the object being scanned) can quantum-tunnel through the gap and be registered on a sensitive crystal in the probe. A significant change in gap width due to the protuberance of an atom will be registered by the probe and translated by software into an atomic contour. It is analogous to a phonograph needle (remember those?) that rides the grooves of a record, reading the glorious music of Mozart in the structure of the grooves.

The scanning tunneling microscope can also pick up individual atoms and deposit them in another location, giving rise to the possibility of constructing molecules according to some functional design, like assembling a model airplane. The new human-made molecule may be a new material of great durability or a virus-destroying drug. Their inventors, Gerd Benning and Heinrich Rohrer, working at an IBM lab in Switzerland, won the 1986 Nobel Prize, and their dream produced a multibillion-dollar industry.

On the horizon and coming soon are two more technologies: nanotechnology and quantum computing. Both are revolutionary. Nanotechnologies, meaning "very, very small technologies," involve the reduction of mechanical engineering, with its motors, sensors, manipulators, and so on, to atomic and molecular scales. Imagine, molecular-scale Lilliputian factories. Compress the scale of any factory a millionfold and you can speed operations a millionfold. This means that quantum-relevant systems for service and manufacture could use the most primitive of raw materials, that is, atoms, and replace our environmentally abusive factories with compact efficient gadgets.

For its part, quantum computing, using quantum mechanical logic systems, could create "an information processing system so powerful that it would be to ordinary digital computing what nuclear energy is to fire."[23]

Chapter 9

GRAVITY AND QUANTUM THEORY: STRINGS

Einstein discovered special relativity. In doing so, he identified the correct *symmetries* of space and time. Prior to special relativity the symmetries of space and time were thought to be *translations* (repositioning a physical system anywhere in space or time) and *rotations* (repositioning a physical system in any orientation). From the work of the famed mathematician Emmy Noether, we learned that these symmetries are connected directly to basic physical principles: the symmetry of translation of a system in time—that the laws of physics are not changing in time—leads to the conservation law of energy. The total energy of an isolated physical system never changes; likewise, the total energy of particles that enter into an interaction is the same as the total energy of all particles emerging from the interaction. (The symmetry of translation in space leads to the conservation of momentum; the rotational symmetry leads to the conservation of angular momentum.) Rotations and translations in space and time are still known today to be valid symmetries for the laws of physics.[1] Einstein, in discovering the principles of special relativity, uncovered the correct symmetry of motion.

Prior to Einstein's theory, in classical physics, there was also a form of "relativity," which is called "Galilean relativity." The "relativity," whether Galilean or Einsteinian, means that physics is the same for all observers in any uniform state of motion.[2] If we do any experiment while traveling on a spaceship near the speed of light, like boiling an egg in water (we assume identical environmental conditions: temperature, pressure, applied heat, etc., as found in a kitchen back on Earth), it will take the same amount of time as on Earth. All the laws of physics that apply to a "stationary system" are the same for a "moving system."

However, Galilean relativity also insisted on a false principle: *time is absolute*. That is, one universal clock could suffice to describe all physics for all observers throughout the whole universe. The perception and measurement of time is not altered by the relative motion between two systems. A key prediction of Galilean relativity follows from this and is the "commonsense" belief that if you chase, with a speed v, after a beam of light that has a speed of c, then you should see the light traveling slower, with speed $c - v$. You should be able to catch up with and overtake a light signal.

Albert A. Michelson and E. W. Morley performed a highly sophisticated experiment (sophisticated for its day in 1887) attempting to find the change in the speed of light as Earth swung about within its orbit around the sun. The experiment yielded a shocking, bizarre, and otherwise confusing, result: the speed of light is constant—it never changes. You cannot overtake a light signal no matter how fast you chase after it! The highway patrol cannot catch, or even close the gap with, a speeder traveling at the speed of light. In fact, the highway patrol cannot travel faster than the speed of light. This discovery set the stage for the other great revolution in physics of the twentieth century, the relativity revolution, in which the key player was Albert Einstein.

The dramatic philosophical change introduced by Einstein was to abandon the absoluteness of time and replace it with a new principle—that the speed of light is a constant for all observers; in other words, the speed of light never changes. The defining principle of the Galilean relativity—the absoluteness of time—that had held sway for three hundred years was therefore discarded. Special relativity, the theory that is built on this hypothesis, leads to profound new consequences about the physics of motion, for example, that moving objects contract in length in the direction of motion, and time slows down for them. Your twin brother taking a trip to Alpha Centauri at nearly the speed of light will return to find you have aged eight years while he has aged a mere two weeks!

Special relativity as a symmetry principle is usually called "Lorentz invariance," in honor of Hendrik A. Lorentz. Lorentz had previously discussed the idea that physical objects were dragged through an ether filling the universe, so that their lengths were contracted in the direction of motion and the mechanisms of clocks slowed down. In this mechanistic view, Lorentz had arrived at the essential relationship of space and time as observed by relatively moving observers, but it was Einstein who straightened out all the logic of the theory and derived the most profound results. Einstein squeezed the Lorentz symmetry out of the sponge of Maxwell's equations for electrodynamics, by invoking his key defining principle that the parameter *c*—the speed of light—would be the same for all observers. So in special relativity, the "absoluteness of time" of the Galilean symmetry is now replaced by the "absoluteness of the speed of light." Moreover, relativity leads to a principle of "causality," by which no signals can travel faster than the speed of light.[3]

Special relativity is therefore completely consistent with the laws of electrodynamics. However, Einstein immediately realized that this required a new theory of gravity to replace that of Newton.

GENERAL RELATIVITY

One of Newton's greatest insights was the "universal law of gravitation." According to Newton, the *magnitude* of the force of gravity exerted on an object of mass by another object of mass *m* is given by the formula

$$F = \frac{G_N mM}{R^2}$$

Here R is the separation between the two objects. This is an example of what is known in physics as an *inverse square law,* that is, a force that falls off in magnitude, or strength, with distance, like (the electric force between two electric charges is also an inverse square law force). The gravitational force law involves a "fundamental constant" G_N. This is called "Newton's gravitational constant" (which explains the subscript "N"). G_N simply calibrates the strength of the force between two masses.[4] Gravity is the weakest known force in nature. To get a feeling for this, pick up a full gallon container of milk. The force your arm is exerting to do this is a little more than eight pounds. This is the approximate strength of the gravitational force of attraction between two completely filled oil tankers ten miles apart.

Newton's theory of gravity cannot be consistent with special relativity. For one, it predicts that the force of gravitation propagates instantaneously between two objects. Newton's theory can only describe slowly moving particles and systems, particles that are *nonrelativistic*. It isn't easy to adapt Newton's theory immediately to special relativity. It ultimately required a shocking and profound new insight into the structure of space and time. With this insight have come revelations that continue to unfold at the heart of modern theoretical physics today.

As we said, it isn't easy to cook up a simple theory of gravity that is consistent with special relativity and Newton's laws in the limit of slow, classical motion. One of the simplest ways to do this would be to propose something like a new "gravitational photon," but it turns out that this would produce repulsive forces (i.e., "antigravity") between equal masses. This would directly contradict experiment, since repulsive gravity is never seen—gravity is always an *attractive* force between any two masses. Another simple hypothesis, called an "elementary scalar field," would produce attractive forces, but these forces would depend in detail on the composition of matter. However, Newton's laws insist that only mass enters the equation. In fact, the "mass" that appears in Newton's gravitational formula is exactly the same "mass" that appears in his famous equation for motion, $F = ma$. This is called the "principle of equivalence," and it was the key that led Einstein in the correct direction of a new theory of gravity.

It took about twelve years from the creation of his theory of special relativity in 1905 for Einstein to write down a full, though nonquantum, theory of gravity (with significant contributions from the great mathematician David Hilbert). This is known as the general theory of relativity, and it is a major intellectual masterpiece. General relativity supersedes special relativity.

At the heart of general relativity, gravitation is interpreted as the bending, or warping, or *curvature* of space and time. In the curved space, particles simply "free-fall" along the best possible approximations to straight-line paths. These paths are the ones with the shortest-distance between two points on the path. Such paths are called "geodesics" in a curved space. For example, the meridian lines on a globe are examples of geodesics in the curved space of the surface of the sphere we call Earth. Airplanes navigate the globe by following geodesics since they are the shortest-distance paths between two airports (latitude lines are *not* geodesics, except at the equator, which is why international flights from New York to Paris do not follow the latitude lines). To find the geodesic between two points on the globe, simply stretch a string or a piece of yarn across the globe's surface from one point to another, such as from Chicago to Tokyo. Over short distances (like Chicago to Des Moines), the yarn makes an approximately straight line, but for the large distances (Chicago to Tokyo or Denmark), it is seen to be a curved geodesic along the surface of the globe.

General relativity thus explains gravitation as a curvature, or bending, or warping, of the geometry of space-time. The curvature of space is produced by the presence of matter, by its mass and energy. Einstein had to teach himself the arcane mathematics that describes the curvature of a space, and after doing so he finally arrived at his new equation for gravity. We can summarize "Einstein's equation of general relativity" for the relationship between curvature of space-time and matter as follows:

$$\text{Curvature} = G_N \text{ times } (\text{ Mass} + \text{Energy })$$

Again we see Newton's original gravitational constant, G_N, appearing in the formula. However, this is a far deeper conceptual formulation than Newton could possibly have guessed.

In general relativity, once the curvature of space (the "left-hand" side of the equation) is established by matter (the "right-hand" side of the equation), objects in motion simply "free-fall" along geodesics through the curved space-time. A space shuttle in orbit about Earth is simply experiencing free fall, where space-time is warped by the presence of Earth, causing the circular orbital motion. Free fall, according to Einstein, is indistinguishable from being out in empty uncurved space, so this produces weightlessness. The curvature of space around the sun, which Einstein's equation predicts, is due to the mass of the sun and produces curved geodesics in space and time in the vicinity of the sun. The planets move on these geodesics in this curved space, essentially freely falling through the curved space. The curvature causes the geodesics to become the elliptical orbits of the planets, with tiny relativistic corrections in Einstein's theory that have been correctly calculated and measured. Planets in orbits, seeming to feel a force of attraction to the sun, are actually in free fall in a curved space-time that is produced by the sun.

Note that curvature of space-time, causing free fall along a geodesic, is a purely geometrical concept—this doesn't involve any "inertial mass," m, of the moving particle as in Newton's formula $F = ma$.

Therefore, the moving particle's mass must—and does—completely cancel out in the formula for the planet's orbit. The "principle of equivalence"—that all objects move the same way due to gravity irrespective of their mass (recall Galileo: a heavy object and a light object fall at the same rate to the ground from the Leaning Tower of Pisa)—happens automatically in general relativity. Another astounding consequence of this, unanticipated by Newton's theory, is that light—made of massless photons—must also move along geodesics. Light is therefore influenced by gravity. That a light beam, grazing our sun on its way to a telescope on Earth, would be deflected by the sun's gravitation was a key prediction of the general theory of relativity.

Newton's theory of gravity is ultimately only an approximation to Einstein's theory in the limit of small velocities of motion compared to the speed of light. General relativity correctly accounts for residual anomalies in the planetary motions, such as the fact that Mercury's perihelion (the location of the distance of closest approach to the sun) advances about one and a half degrees per century, an effect for which Newton's theory cannot account. General relativity also correctly predicted the bending, "lensing," and color shifting of starlight as it passes or leaves gravitating objects. Einstein's general theory of relativity applies to the universe as a whole and correctly predicts that it should be expanding, that space is being created. The crucial prediction of Einstein's theory—that starlight, traveling on a geodesic, would be bent by the sun—was confirmed by observations of solar eclipses in 1919.[5] This observation established general relativity as scientific fact and catapulted the obscure Albert Einstein into worldwide fame as the superstar of science.

As we'll now see, general relativity predicts that objects can become so massive that they can trap all matter and light from ever escaping from their surfaces.

BLACK HOLES

Let's ask a simple question: "What happens if a particle attempts to escape the surface of a planet that has such a strong pull of gravity that it would require *all the particle's rest energy*, $E = mc^2$, in order to escape?" Indeed, the massive planet would then forbid the escape of the hapless particle, since there would be nothing left of the escaping particle. All its mass would be expended in the escape process. Even light, a photon, could not escape, since no finite energy would be left to the photon once it escaped.

Such a massive object is called a *black hole*. Any planet can be a black hole if we squeeze all its mass down inside a sufficiently small sphere with a radius R, called the *Schwarzschild radius*. Any object of a given mass M, whose radius is smaller than R, as determined from the formula, $R = 2G_N M/c^2$, will become a black hole. Nothing can escape a black hole from any distance within the Schwarzschild radius.[6] Fortunately, Earth is far from being a black hole. If Earth were the massive *object*, and we put in numbers, we would find that for Earth to become a black hole, it would have to be compacted down to the tiny radius of meters, or about a quarter of an inch! For the sun, the Schwarzschild radius is about two miles, so if we compressed the entire sun down to the size of a small town, it would become a black hole! The density of the matter of the sun filling a region this size would grossly exceed that of the atomic nucleus. Nonetheless, today it is widely believed by astronomers that the centers of most galaxies contain humongous black holes that have masses many billions of times greater than the mass of the sun.

The Schwarzschild radius is not necessarily the surface of a black hole but rather a distance from the center at which there is an "event horizon"—the place from which light can no longer escape. Aim a powerful beam of light outward, and it will lose all its energy and no light will get away from the event horizon. When objects fall into a black hole, they cross through the event horizon.

Paradoxically, a stationary observer external to the hole, who must be constantly accelerating away from the hole to avoid falling in, never sees the object actually cross the event horizon. From his perspective, it takes an infinite amount of time. The external observer witnesses time frozen on the horizon, but nothing able to escape. Any light the objects on the horizon, would emit would be "redshifted" toward the far infrared (lose energy) as it tries to escape, ultimately to zero energy. To the outside observer (us!) the objects would simply fade eternally into the event horizon, which would become darker and darker. In reality, astrophysical black holes at the centers of galaxies are surrounded by things falling in—enormous gas clouds and whole star systems. These create vast amounts of high-energy radiation as they collide and accelerate, before they ever hit the horizon. So we never really get to see the black hole horizon, the great pool of unshrouded darkness.

> Turning and turning in the widening gyre
> The falcon cannot hear the falconer;
> Things fall apart; the centre cannot hold;
> Mere anarchy is loosed upon the world,
> The blood-dimmed tide is loosed, and everywhere
> The ceremony of innocence is drowned;
> The best lack all conviction, while the worst
> Are full of passionate intensity.
>
> William Butler Yeats, "The Second Coming"[7]

If, on the other hand, you were falling into the idealized black hole, one without a lot of nearby orbiting debris, you would notice nothing (well, there would be extremely large "tidal forces" that would rip you apart, but these would be vanishingly small for you if you were a pointlike particle). You would cross the horizon in very little time and would continue your free fall inside the horizon toward a compact, dense inner "singularity"—all that is left of the squeezed matter within the hole. Scientists don't know what the laws of physics would be at the singularity, but it could be something like string theory (more on that later).

The fact that in-falling matter can cross the horizon, while stationary external observers never see anything crossing the horizon in any finite time, is what characterizes or defines a "horizon." Space and time are decisively sliced into two distinct regions by a horizon—the outside of the black hole and the inside—and ne'er can the two communicate. You can only take a one-way trip into a black hole.

QUANTUM GRAVITY?

Physicists began to seriously contemplate quantum mechanics together with gravity in the 1950s, and they immediately encountered problems. The biggest headache was that the infinities, which we saw in the case of the vacuum energy of the Dirac sea in the previous chapter, now returned with a vengeance. A theory of quantum gravity simply made no sense—almost everything physicists tried to compute was plagued with infinities. Everything was now incalculable, and the theory was useless.

The first issue that arises when quantum effects are blended with gravitation was first noted long ago by Max Planck himself. From Newton's fundamental constant of gravity, G_N, combined with the speed of light and Planck's constant, we can mathematically construct a "length scale." This length scale is called the "Planck length," called L_P ("ell-sub-P") and it is given by the formula

$$L_P = \sqrt{\frac{\hbar G_N}{c^3}} = 1.6163 \times 10^{-35} \text{ meters}$$

Notice the appearance of the three famous "fundamental constants" of physics in this one formula, $\hbar$, c, and G_N. The Planck length, about 10^{-35} *meters*, is miniscule, even when compared to the size of an atomic orbital, 10^{-10} meters, or to an atomic nucleus, 10^{-15}, or to the shortest distances we have yet probed with particle accelerators, 10^{-18} meters. At the ultra-tiny Planck length scale we know that gravity can no longer be approximated as a nonquantum (classical) phenomenon. At this scale, space and time as classical entities must jiggle and fluctuate wildly. Space and time would become "fuzzy" or, as some have proposed, become a "space-time foam"—a broiling, bubbling quantum chaos. A theory of quantum gravity must tell us in detail what is happening deep down at the short distance of the Planck length. And, it must instruct us as to what is happening there.

It was well known that there are waves that can propagate in the classical version of Einstein's general relativity, just like Maxwell's description of light emerged as a wave of electric and magnetic fields. In general relativity these are called *gravitational waves.* Just as Maxwell's classical light waves were later found to be quanta—called photons—no one would doubt that gravitational waves are also comprised of quanta, which we call "gravitons." Like photons, gravitons would be bosons. Gravitons in general relativity are also found to be spin = 2 particles (whereas the photon has spin = 1 and the electron, a fermion, has spin = 1/2).

To this day, we have never detected a graviton in any experimental detector. In fact, we have never even directly detected gravitational radiation of any kind. We indirectly know it exists, because certain astrophysical systems (binary pulsars) are slowing down as their orbits decay, and this matches the expected rates from their energy being carried away by gravitational radiation.[8] The problem is that gravity, as we have already noted, is a very feeble force. There are ambitious experiments attempting to detect the gravitational radiation believed to be radiated by large astronomical systems. This would detect "classical radiation," that is, many trillions of gravitons in a single wave, but even if we did, we would still be very far from observing a single quantum of gravity—a graviton.

But the real problems for a quantum theory of gravity begin when we start considering the self-interactions among gravitons; for instance, how does a graviton itself emit and absorb gravitons? We encounter a morass of mathematical problems that demand a clear definition of the theory at the very shortest distance scale, the Planck length.

STRING THEORY

Largely out of the need to have a consistent theory of quantum gravity, there has emerged an entirely new set of ideas of what all particles in nature are. This new paradigm is called *string theory*.

We have talked of "particles" throughout this book without really defining what we mean by a "particle." Physicists always make simplifying approximations. For them, a particle, in its simplest form, is just a pinpoint in space where some mass is located. So to describe a particle we need only say where it is at any given time, that is, to introduce a "trajectory" $x(t)$ together with the mass m.

If we were following Newton, solving his equations for the motion of a planet about a star, we could treat both the star and the planet as pointlike particles. Of course, particles in reality are more complicated. An atom, for example, is only a particle in the approximation of looking at it through a low magnification microscope. If we crank up the magnifying power we will eventually see the electrons in their cloudlike motion around the nucleus. The nucleus will then look like a pinpoint particle of mass at this scale. But if we go up further in magnifying power, by another factor of a hundred thousand, we would see that the nucleus contains protons and neutrons. And the protons themselves appear as particles on the nuclear scale but are seen to contain *quarks* at much shorter distance scales.

Quarks and *leptons* (the electron is an example of a lepton) appear to us to be truly elementary particles. In addition to their tiny masses, they also carry other attributes, such as spin, electric charges, and quark color charges. Yet even quarks and leptons have no discernable size. These particles are true pinpoints of matter as far as anyone can tell by any experiment to date. Suppose, however, we could increase the magnifying power of our microscopes (the energies of our particle accelerators) by many trillions of times. Would a quark or a lepton still appear as a pinpoint? Or would we begin to see a fuzziness, perhaps a cloudy shape of something inside, resembling a "hyper-atom" with a "hyper-nucleus" and "hyper-electrons" orbiting around deep inside the quarks and leptons?

String theory begins with the notion that the fundamental objects in nature are not particles, that is, they are not pinpoints of matter, and that the basic conceptual starting point of a pointlike particle is simply wrong. If we ask the purely mathematical question: "What's the next most complicated thing beyond a pinpoint particle?" we can give an answer: the next most complicated thing is a *string*.

A particle is described as a pinpoint of matter located somewhere in space. We describe its trajectory by giving its position as a function of time, $x(t)$, and this traces out a path or trajectory in space-time. We plot this in a space-time figure, such as figure 35. The path of a particle is called a "world-line," and it traces out the motion of the pinpoint particle in time, as shown on the left. If we take a snapshot at any time, all we see is the location of the dimensionless point in space.

Now consider a string. A string is a *one-dimensional object* instead of a point. A string sweeps out a "ribbon" in space-time, as seen below in figure 35. In a snapshot of a string, that is, for a fixed value of time (slicing through the string for a fixed value of t), we see the string indeed appears as a little extended object in space. The string has a size, that is, a length, in one dimension for a fixed time, t.

The position of a point on a string, at any fixed time, can be defined by a position, $x(y)$, where y is an "internal coordinate" on the string, as in figure 36. At $y = 0$ we have one endpoint of the string, and at $y = L$ we have the other. This is called an "open string." Or else, the string could be a closed loop, where $y = 0$ and $y = L$ are the same points, $x(0) = x(L)$. This is called a "closed string." The string then can be specified at any time, t, by giving a mathematical function of time, t, and y, that is, $x(t, y)$. If we plot the motion of an open string in space and time, we see it sweeps out the ribbon. For the closed string, we see a kind of tube that moves in space-time. The ribbon for the open string, or the tube for the closed string, is called the "world-sheet" of the string.

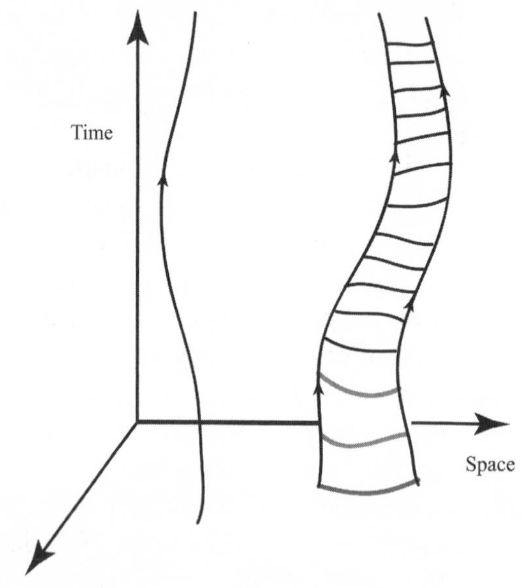

FIGURE 35: A "particle" is a "world-line" that runs continuously from the past into the future. At any instant the particle is just a pinpoint in space. A string is a "ribbon" or "world-sheet" that runs continuously from the past into the future. At any instant in time, the string is a one-dimensional object.

Newton's equations determined the motion of particles. In relativity these become space-time equations that determine world-lines. These are prequantum descriptions of things. The effects of forces cause the particle to be deflected from the geodesic path. Now, however, we have the extended object in the form of a string—so what is the new principle?

For strings, the principle governing their motion is that the "world-sheet" or "ribbon" swept out in time has *the least possible area*. The area of the ribbon is a bit like the shape that a soap film takes when soap is attached to a bent wire. In a prequantum (or classical) string theory, we discover that the string will lose all its energy and will then always collapse down to a point. Recall that this was the problem faced by Bohr and others with the prequantum hydrogen atom, which collapsed down into a dead object by radiating all its energy away. However, just as quantum theory saved hydrogen, it also saves string theory. Quantum strings move through space and oscillate and wiggle, and how they oscillate determines what particle they appear to be when seen in an accelerator. All the myriad particles, the quarks, leptons, photons (and other "gauge particles), and even gravitons are simply particular vibrations of one and the same kind of string.

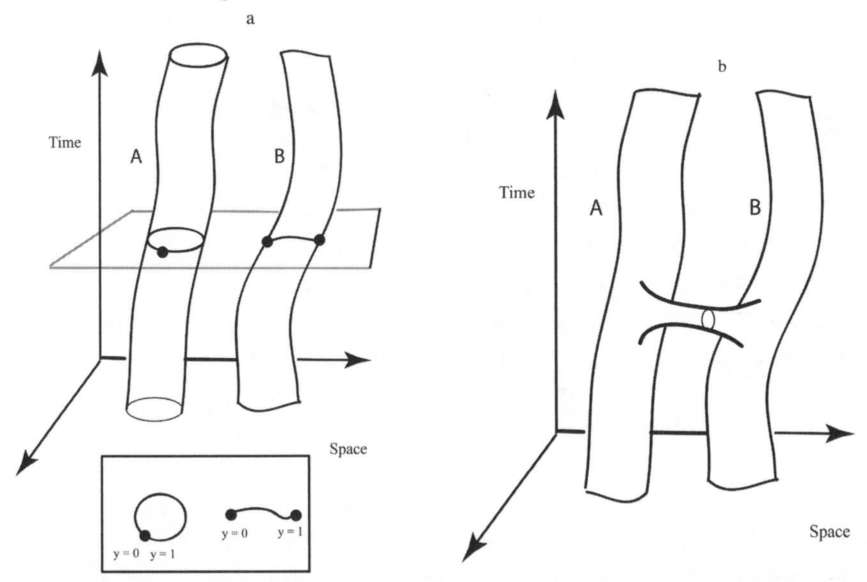

FIGURES 36 a and b: Strings can be either "closed" or "open." All string theories contain closed strings, but open string theories are optional (requiring something to define the edges). At any instant in time, the open string is a line segment and the closed string is a circular loop. In figure 36b we see that string interactions are just a deformation of the string world-sheet, as though it is made of rubber, connecting together adjacent strings. All string theories have an interaction where a closed string connects to two adjacent strings. This yields gravity in string theory.

If we consider the world-sheet of a string moving through space and time, it looks like a ribbon in space-time. We can try to "decorate" the sheet by drawing a map or internal "string coordinates" on the ribbon. A new symmetry now arises in string theory which states that: the physics cannot possibly depend on what choice we make for these internal coordinates. This is sort of obvious—we humans only introduce the coordinates as a computational device for the area of the ribbon. But we must be very careful that this symmetry is also preserved by the quantum theory and that by drawing the coordinates we aren't somehow redefining what we mean by the string. This symmetry is called "Weyl invariance," for the great theorist of the early twentieth century, Hermann Weyl, who discovered it in other contexts. It means that there are no mile markers and no highway maps on the string ribbon . . . nada . . . zilch! Put another way, any map we make of the surface of the string world-sheet in space-time should be as good as any other because there are no special features on the string ribbon. This is the content of Weyl symmetry.

So, what happens if we insist on Weyl symmetry in the quantum theory of strings? We find something quite remarkable. The number of dimensions of space and time is constrained to certain special values. In the simplest string theories, a so-called bosonic string, where we do not have fermions, we get D = 26. That's twenty-five space dimensions plus one time dimension, like saying we live in a universe with twenty-five dimensions of space. In other words, bosonic strings can only be consistent when they exist in twenty-five space dimensions plus one time dimension. However, that's not exactly our observed world.

Theorists then introduced spinors (spin-1/2 particles) onto the world-sheet of the string (recall, spinors are square roots of vectors, and these are just square roots of coordinates). The first such fermionic string scheme was written down in the early 1970s by Pierre Ramond and independently by John Schwarz and Andre Neveu. In addition, Ramond discovered that the string theories now contained a new symmetry, the one we encountered in chapter 8, that has become a favorite of theoretical physicists ever since: supersymmetry.

SUPERSTRING THEORY

Armed with fermions on the string, when one repeats the test of Weyl symmetry, one now sees that there has been progress—the dimension of space-time is reduced to D = 10; that is, we now predict nine space dimensions plus one time dimension. Essentially, it is the exact cancellation of the vacuum energy on the string that is essential to maintain the Weyl symmetry, and this requires the supersymmetry. Supersymmetry is now predicted to exist in nature if we believe in string theory. But we are still far from the three dimensions of space and one dimension of time that we see in our universe.

All this was done at a time when practically no one was paying any attention to strings or string theory, in the 1970s. It was an exciting time for experiments and theory, but strings and even supersymmetry were just not that interesting back then. People were not centrally focused at that time on the problems of quantum gravity.

In the mid-1970s, during a visit to Caltech, Joel Sherk, a young and brilliant French theorist, teamed up with John Schwarz. They had noticed an interesting feature of string theory: all string theories, whether open or closed, with or without fermions, contain a common mode of vibration (like our guitar string modes back in chapter 6; a mode of vibration is just what you get when you pluck a violin or guitar string and it then vibrates at some frequency). In this case, Sherk and Schwarz observed that this particular mode behaved like a massless "particle," and it had spin-2. Recall that the graviton, the quantum of gravitational radiation, is massless and also has spin-2. This mode of oscillation, common to all strings, is exactly like a graviton. Hence Sherk and Schwarz wondered if string theory might offer a solution to the outstanding problem of providing a sensible quantum theory of gravity. The problem remained, however, how do you get from ten space-time dimensions down to the four that we observe?

In 1974 Sherk and Schwarz proposed that the six extra dimensions (we see only four dimentions of space-time, while the fermionic superstrings require ten) predicted by string theory are rolled up into a small ball, or "compactified" so small that we don't detect them with our meager low-energy particle accelerators. All we see at low energies are the residual four dimensions that extend throughout our universe. This has the added virtue of converting various string vibrations into things that act like the other observed particles that could in principle explain the known list of particles in the Standard Model. The idea provides a rationale for all the observed forces in nature, all coming from one common stringy source that also yields gravity.

Yet string theory still remained an off-the-beaten path and arcane hypothesis. It was mainly practiced at Caltech with a few smaller efforts around the world. Supersymmetry, sans string theory, however, had come alive in the mainstream theory community and was rapidly becoming a potential master theory of all forces, with gravity essentially left on the margins. But soon theorists began to consider more seriously if all this could somehow be built into one master string theory. Bear in mind, there is to this date no evidence for the existence of strings or supersymmetry in nature—but there is no counterevidence either. The most compelling thing is that gravity forms the centerpiece of string theory, and we had no other successful quantum theory of gravity.

However, another potential mathematical disaster occurs when string theory is adapted to include all the (nongravitational) forces in nature in the low-energy theory. It is called a *gravitational anomaly*, and it represents a complete breakdown of Einstein's general relativity. That is, if there is a gravitational anomaly, Einstein's basic equation where matter generates curvature could not be correct—the matter side of it, that is, the right-hand side of the equation, which contains the nongravitational string vibrations, is what goes haywire here. In short, the gravitational anomaly means there is no way that curvature (the left-hand side of Einstein's basic equation for general relativity, where the curvature controls planetary orbits), can be set equal to matter (the right-hand side, where the mass of the sun would appear). This might have been the final death blow to nascent string theory; still, string theory remained on the periphery of theoretical physics.

However, in 1984, John Schwarz and Michael Green faced the issue of the gravitational anomaly head-on. They carried out a tedious calculation that approached completion during a violent thunderstorm over Pasadena in their Caltech office. The calculation showed that only in very exceptional circumstances, there is no gravitational anomaly. String theory can work only in these special cases. Schwarz and Green published the results of their heroic calculations of the gravitational anomalies. Strings could now be compactified down to the four dimensions of space and time, and string theory now uniquely predicted that the other symmetries describing all other forces in nature beyond gravity were of a special form (this has a "symmetry" called "$E_8 \times E_8$"). The Standard Model fits comfortably into this framework.

The Green-Schwarz calculation unleashed a veritable flood of intellectual fervor. It suddenly became the height of respectability to be a string theorist, thus multitudinous scientific papers on strings flowed. New answers to old questions began to emerge, topology began to play a fundamental role, and deep ideas on why quantum theory exists began to surface. This is a big topic, and we cannot go into it within the confines of this book; however, a lot has been written about it, and we heartily recommend Brian Greene's book *The Elegant Universe*.[9]

At the time that string theory took off, many theorists were completely sold on the new "theory of everything." Some physicists proclaimed that this was the greatest of all scientific revolutions and that more physics had been done in creating string theory than in the entire preceding century (more so than when Bohr solved the hydrogen atom or Schrödinger proposed his equation!). So how does that square with today's view?

STRINGS TODAY

String theory has had profound impact on our understanding of the quantum science of gravity and associated theoretical physics. Nothing has come closer to realizing an elegant unified theory of all the forces. String theory raises foundational issues about the basis of quantum theory as well.

String theory achieves at least two things: it postdicts for us a quantum theory of gravity and it predicts that supersymmetry will eventually be demonstrated to exist in particle accelerator experiments. Indeed, supersymmetry may be detected in experiments within the next few remaining years of the Fermilab Tevatron and as the Large Hadron Collider (LHC) ramps up at CERN.

Unfortunately, at this time, no experimental evidence for any of this has emerged. Of course, it isn't necessary that supersymmetry (SUSY) be seen at the LHC; string theory can still be true with SUSY showing up at still higher energy scales. One may have to wait for even more energetic accelerators, or it may even be well beyond and out of the reach of experiment.

In the meantime, the fine art of string theory has diverged considerably from the frantic reductionism of 1984. Soon theoretical physicists recognized that there are many ways to descend from ten to four dimensions of space-time (one famous model has parts of the string living in twenty-six dimensions and others in ten dimensions). Then it was recognized that the "modes" of strings include other objects, called *branes*, which can be viewed as just as fundamental as the strings themselves (branes are not just one-dimensional objects, but are bulkier multidimensional objects; for example, ordinary space itself can be thought of as a three-dimensional brane). This has provided further enrichment of the subject and has opened the doors of thought to further theoretical vistas.

One striking insight came in 1996 when Juan Maldacena at Princeton examined a particular dimensional compactification step from five dimensions to four dimensions. The five-dimensional world is something called an "anti-deSitter space," or "AdS" for short. This is a space and time that is very highly warped. The four-dimensional world is a slice through this space by a four-dimensional surface called a brane. Maldacena compared how nonstringy quantum physics on the brane could be related to the string physics in the five-dimensional AdS space. He found that a particular theory on the brane (a quantum field theory called "N = 4 super-Yang-Mills") has the same behavior as the string theory in the five-dimensional AdS space. In other words, an observer in the five-dimensional world would see a tangle of oscillating and interacting strings, while an observer on the brane would see a world of many particles—and the physics of these two worlds is the same!

This famous result is called the "Maldacena conjecture" and is a precise theoretical example of holography, where a higher-dimensional theory is described by the boundary one dimension lower. Recall that this is the idea we encountered in the previous chapter to solve the vacuum energy problem of the Dirac sea, whereby the entire physics of the volume of the Universe can be holographically represented by the activities on a lower-dimensional boundary. This strongly encourages us to think that we are missing something in how we compute vacuum energy—or conversely, that something in our description of space and time is expendable.

Holography is a work in progress, but it looks promising for future research insights and may ultimately reshape what we think quantum physics represents about nature.

One very pronounced difference between the revolution of string theory and the development of quantum physics in the early twentieth century is the virtually complete lack of any experimental input into string theory. This is no fault of string theory or string theorists—it is simply the way things are. Strings, with their internal substructure, are too far removed into the shortest-distance scales of nature to be accessible to particle accelerators. Our best hope to begin to verify them, in principle, is to discover supersymmetry in our particle accelerators, but that has yet to occur. Nonetheless, this raises a question for you to ponder: Could Heisenberg, Bohr, Planck, Schrödinger, Einstein, Dirac, Pauli, and others ever have created the quantum theory without the experiments that forced them to do so and that guided their thinking? To us, it would seem incredible that, after retracing the steps of these early twentieth-century masters, who lived in a time of countless new experiments connected to their scientific inquiries, anyone could ever distill everything about the world from pure thought. The closest anyone seems to have gotten to achieving this was Einstein, who lived in a world of fast and furious experimental results. However, even the great Einstein made the error of ultimately rejecting the quantum theory.

One might reflect upon how anyone could have guessed that electrons have spin-1/2 and are described by spinors, the square roots of vectors, and that quantum theory is a square root of probability. Clearly the early twentieth-century pioneers needed the experimental weight of atoms, the blackbody problem, and the photoelectric effect to direct their thinking. Would Dirac ever have been compelled to write down his equation and his "Dirac sea," which led to the prediction of antimatter, without the experimental success of Pauli's exclusion principle, which gave us the tangible understanding of the periodic table of the elements?

Nevertheless, one could argue (and it is argued) that we now have most of the ingredients in hand needed to complete our understanding of nature. The world of physics is much different now than it was in the early twentieth century. In the era of the quantum revolution, most things were not understood, but theory teamed up with experiment to provide a new clarity. Today, we know of no experimental result that is inconsistent with our "Standard Model" of particle physics, yet we also know that our Standard Model is incomplete. Some argue that it's just a matter of following mathematics wherever it takes us. Perhaps. But ultimately it isn't a matter of whether or not the theorists have done their mathematics correctly—it's a matter of what the limited imagination of humans may have left out.

THE LANDSCAPE

We'll end with what may be the endpoint of string theory, if not of all theories themselves. It is an idea that has taken hold and was advanced largely by Leonard Susskind of Stanford.[10] It is, perhaps, one of the most profoundly illuminating statements we can make about nature, and one of the most sobering.

Suppose you want to predict exactly what the laws of physics are as seen in low-energy experiments (that would include our best accelerators, like the LHC) from string theory. Is it possible? The problem is one of determining the vacuum state in string theory. For any quantum theory, the first order of business is "what is the ground state" or, equivalently, "what is the vacuum?" The vacuum is the analog of the hydrogen atom in a periodic table of elements—it is the starting point of understanding all the other atoms.

Various modes of vibration of the string, called "moduli fields," condense into a kind of quantum soup in the vacuum. The values of the moduli fields determine the "laws of physics," that is, the values of such things as the electron mass or Newton's gravitational constant, in any particular region of space. The moduli fields can vary slowly as we traverse vast distances of space, so the laws of physics can be effectively different elsewhere where the moduli fields take on values different from what they are here.

Theorists, such as Michael Douglas of Rutgers University, have estimated the number of possible vacuum states that can occur in string theories due to the various values of the moduli fields throughout space. The answer is about 10^{450}. That is a hell of a lot of vacuum states! One might think nature somehow flips a coin and chooses one of these vacuum states, and that's where we end up.

But now consider the coincidences that have to happen for us to merely exist. For one, we live in a very large universe with a very small cosmological constant (vacuum energy density), and this is a good thing because a small universe might be too dense (too hot!) and might not be around long enough for evolution to occur, and too large a universe might make it unlikely for enough concentrated matter to clump to form a solar system (too cold!), and so on. The cosmological constant drives the expansion of the universe and is measured to be quite small. This appears to be a complete accident, since no one has come remotely close to a decent theory of it; that is, no one has any idea how to compute the cosmological constant. Other coincidences have to do with the strengths of forces in nature that permit the synthesis of carbon within stars, necessary for life. Why would just one universe exist where these coincidences are dialed in so precisely so that we can enjoy good health, beaches, and shrimp?

Susskind and others have advanced the intriguing idea that the universe we see is a tiny, miniscule, insignificant fraction of an enormous super-universe in which every other vacuum state also occurs, way out there, somewhere. We can't see these other universes because they simply reside way beyond the horizon of our own universe. The maximum distance we can ever see, our horizon, is defined by the farthest distance that light has traveled since the big bang origin of the mega-universe, and that is a mere thirteen billion light-years. The metaphor for this is that we are confined by our horizon to live on a quarter-sized patch in a farm field somewhere in Kansas, and the whole surface of Earth is the mega-verse, rich with mountains and seas, glaciers and jungles. (In fact, even this metaphorical comparison is off by several hundred orders of magnitude).

This grand mega-verse is called the "Landscape." We exist, so we must be in a portion of the Landscape that is habitable, so the marvelous coincidences have randomly happened. If they aren't the same elsewhere, so what? That's not where we live, or where anything lives, for that matter. So perhaps no life exists in the Landscape's remote and unknown "mountain peaks" or the deepest depths of its "oceans." The Landscape offers a rationale for the "anthropic principle," a virtual tautology that maintains that what we see must be fairly ideal to exist because we happen to exist. It's a bit Panglossian. Indeed, it's hard to think about the Landscape and even harder to write about it. We recommend that you take a look at Leonard Susskind's book, *The Cosmic Landscape: String Theory and the Illusion of Intelligent Design*.[11]

Whether or not we accept the Landscape, there remains a real and sobering aspect to this. The universe indeed has a finite, observable size. Given its age and the speed of light, there is a horizon of the universe beyond which we will never see—some thirteen billion light-years, in any direction. Any physics that affects the larger scales in nature cannot be tested with any good statistics by the observations limited to such a small world. Given our position in the universe, we may never see enough of the universe to figure it all out.

> see a World in a Grain of Sand
> And a Heaven in a Wild Flower,
> Hold Infinity in the palm of your hand
> And Eternity in an hour.
>
> From William Blake, "Auguries of Innocence"

Chapter 10

QUANTUM PHYSICS FOR MILLENNIUM III

As we've seen throughout this book, quantum science, despite its altered reality, actually works—in fact, it works miracles! successes are dazzling, profound, and far reaching. It has led to our understanding and control of processes involving molecules, atoms, the nucleus, and subnuclear particles, together with the forces and novel laws that explain the micro-world. The deep intellectual discourse of the early twentieth-century quantum founders has given way to a powerful tool that, today, allows us to engineer the astonishing gadgets that are reshaping our existence.

Out of quantum wizardry have come technological instruments of undreamt-of power—from lasers to scanning tunneling microscopes. Yet a few of the intellectual giants who have created quantum science, written the textbooks, and devised the miraculous inventions still lie awake at night and worry. They worry because of the niggling suspicion that, as Einstein pointed out in his papers, quantum science, for all its glitter, may not be the whole story. How can probability truly be a part of nature's fundamental principles? There may be something left out, something missing. In fact, gravity is an example of something left out of quantum science for many years, and the dream of a consistent theory unifying Einstein's general relativity and quantum mechanics inspired some fearless theorists to work at a fundamental level, where only abstract mathematics provides a lantern, to create string theory. But is there something deeper, a missing component to the logical structure of quantum theory? Like trying to solve a master jigsaw puzzle at times of frustration, we wonder, Did a key piece fall out of the box?

Everyone's overriding hope is that we may soon discover an even more powerful super-theory that contains quantum theory in a certain limit, like relativity devours Newton's classical mechanics, the latter restricted to the limit of things that move slowly. This would mean that contemporary quantum theory is not the end of the line, but that far, far out there in the mind of Nature there is a final theory, a better, more inclusive description of the universe. This ultimate theory would not only address the frontiers of high-energy physics, molecular biology, and complexity theory; it could also lead us to entirely new phenomena that have so far eluded researchers. We're a curious species, after all. How can we resist probing this quantum world, which is as exciting and surprising as a newly discovered planet orbiting a distant star? It's a serious business, too, if 60 percent of our GDP depends on the mastery of quantum science. For all these reasons it is important to keep investigating the basic structure required to understand nature.

"Quantum phenomena challenge our primitive understanding of reality; they force us to reexamine what the concept of existence means," writes E. J. Squires in the preface to *The Mystery of the Quantum World*. "These things are important, because our belief about 'what is' must affect how we see our place within it; and our belief in what we are. In turn, what we believe we are ultimately affects what we actually are, and how we behave."[1] Heinz Pagels, the late theoretical physicist and author of *Cosmic Code: Quantum Physics as the Language of Nature*, describes the situation as akin to a large mall with a variety of shops, all selling "reality" in various forms.[2]

We challenged our notions of reality when we visited Bell's theorem and its experimental consequences in the previous chapters. Recall how we were forced to consider the possibility of nonlocal effects: instantaneous transfer of some influence between two detectors that may be any distance apart? Here one gets the classical impression that the measurement at one distant detector is influencing the measurement at the other. The only link between the detectors is the pair of particles (photons, electrons, neutrons, and so on) that were "entangled" in their quantum state at birth in the source and that subsequently arrived at the two detectors, 1 and 2. If detector 1 determines that its particle has attribute A, then detector 2 must measure attribute B for its particle, or vice versa. In the view of quantum wave functions the act of measurement at detector 2 is "collapsing" the quantum state throughout all of space instantaneously. This was abhorrent to Einstein, who held to a belief in locality and the fact that no signals can travel faster than the speed of light. The experiments have ruled out any influence other than the act of detection at 1 and 2; in other words, we can eliminate the possibility that the arrangement of the detectors at 1 has somehow communicated to the detector at 2. But the existence of entanglement is a fact and is confirmed by the experiments—quantum theory is again shown to be foundationally correct. The devil is in our own reaction to confronting this new, seemingly paradoxical reality. As one theorist put it, we really want to feel a "peaceful coexistence" between quantum mechanics and relativity (breaking the cosmic speed limit is only for daredevils).

The central question is whether the Einstein-Podolsky-Rosen issue is just an illusion, and perhaps phrased in a way that makes it seem counterintuitive? Feynman himself was engaged by Bell's theorem and tried to find a better description of the quantum theory that could make its reality more palatable, even though his own sum over paths comes close. As we saw, he expanded some ideas of Dirac and invented another way to think about quantum physics, called the "path integral," or the "sum over histories." In this picture, when a radioactive particle decays into a pair of particles, one spin-up and the other spin-down, there are two "paths" throughout space-time to consider. One path (call it "A") delivers the spin-up particle to detector 1 and the spin-down particle to detector 2; the other path (call it "B") delivers spin-down to 1 and spin-up to 2. Each path has a certain "quantum amplitude," and we add up the amplitudes. When we make a measurement at detector 1, we are discovering which of the two paths the system took, that is, if we measure spin-up at detector 1, the universe took the first path "A." All we can compute is the probability (the square of the amplitude) for any given path.

In this "space-time" picture, gone is the idea of instantaneous propagation of information across light-years of space. The situation is more akin to the classical one in which our friend has sent us and our colleague, who is on Rigel 3, one of two colored billiard balls (red or blue), and we find that we received the blue ball. Therefore, we know instantly that our pal received the red ball. Yet nothing changes throughout the entire universe; we simply learn which of all possible options actually happened. Perhaps this lessens the philosophical angst over an EPR experiment, but it is not to say that the idea of a quantum sum of paths comprising reality isn't astonishing in itself. We can delve into why the path integral works, and indeed it works so as to prevent signals from being transmitted faster than light—in fact, this is intimately related to the existence and properties of antimatter and quantum field theory (as we've seen in chapter 8). Here we see that the whole universe is governed by an infinite set of possible paths that govern its evolution in time. The whole universe moves forward in time like a vast wave-front of possibilities. Only occasionally do we measure which path by doing an experiment at some event in space and time. The wave regroups and continues onward into the future.

These issues have caused a generation of physicists to cry out in frustration about what quantum physics really is. To this day, the conflict with all intuition and experience with the reality of quantum mechanics is confounding.

> Anyone who is not bothered by Bell's Theorem has to have rocks in his head.
>
> David Mermin[3]

But, ultimately we need to accept it:

> So irrelevant is the philosophy of quantum mechanics to its use that one begins to suspect that all the deep questions are really empty. . . .
>
> Steven Weinberg[4]

Weinberg's comment is not dismissive; rather, it is profound. It isn't clear that anything we have to say about the deep philosophical meaning and interpretation of quantum theory is going to teach us anything. Scientifically, quantum theory simply works, whether or not we philosophically understand it. Bell's theorem teases us with something that is fundamental to quantum theory, the mixed and/or entangled states. Yet these things occur throughout all the phenomena of the physical world, implicit in the structure of a benzene molecule, a K-meson, or the vacuum state of the universe. It's a part of a greater whole.

Nevertheless, some physicists rolled up their sleeves, and by the end of the twentieth century they had managed to carry out various exquisite precision experiments on the foundations of quantum science. Did these experiments deliver complete answers to us? Not at all, but they have served to sharpen intuitions in a domain where much is counterintuitive. And, to the surprise of the physicists on the sidelines watching the developments with bemusement, these bizarre ideas appear to have (gasp!) practical uses. Quantum uncertainty and quantum entanglement are the parents of quantum cryptography! And from spooky action at a distance, that is, nonlocality, we may eventually derive the wonders of ultra-high-speed quantum computation. And these disquieting, seemingly nonlocal effects had the goal of goading some theoretical physicists to devise (sometimes desperately) alternative approaches to understanding quantum mechanics.

SO MANY WORLDS ... SO LITTLE TIME

Recall that in the "Copenhagen interpretation of quantum mechanics," the particle may as well not exist until the act of observation. The act of measurement forces the particle to be in a definite quantum state with definite properties. It causes the wave function to collapse, from its noncommittal list of possibilities, each with an associated probability, into a single certain state—the result of the measurement. The conceptual difficulty here is the importance of the observer, a tinge of subjectivity that makes scientists very uncomfortable. After all, the universe got along very well without observers for ten billion years or so, as far as we know. Why should we suddenly require them? Also, how does the act of observation cause the wave function to collapse?

An alternative to this interpretation was put forward in 1957 by a Princeton graduate student, Hugh Everett. Everett's audacious suggestion (modified some years later by Bryce de Witt of the University of Texas) was that the particle exists and is in all the possible states embodied in the wave function. But now, each of the possibilities exists in a different universe. Thus, if a photon heads toward a barrier, such as the window at Victoria's Secret, the entire universe splits into two universes. In one universe, the photon penetrates the barrier; in the other, it is reflected. The observer, and everyone and everything, also splits in two, each universe following one of the photons. Thus we have, from this single event, two universes. (Contrast this to Feynman's path integral where the observation simply plucks out the one path the universe took containing the two possibilities.)[5]

Obviously, according to this scheme, we exist in any moment in an infinite number of universes of which we are unaware, with a like number of observers who are similarly oblivious to one another. Could an observer in one parallel universe carry on a romantic affair with someone in another? This is a high price to pay for eliminating the awkwardness of influential observers and collapsing wave functions. The "many worlds" are actualizing the many paths. Bohr might not be so sympathetic: it's endowing some reality to things we don't measure.

When we have a quantum system—a particle in a magnetic field, say, an electron entering a magnet—we describe the system by a Schrödinger equation and we come up with a number of possibilities for the results of future measurements. For example, the energy levels of the atom may have five or seven possibilities; the deflection of the electron spin may be "up" or "down." Each possibility has a probability. When we actually make the measurement, conventional quantum science then says that the (probabilistic) wave function "collapses," and suddenly the energy state is definite, for example, with 6.324 eV of energy or the spin is definitely "up." This gives us two kinds of unwanted intellectual baggage: it requires the action of the observer and the concept of the collapse of the wave function. Everett's interpretation implies that nothing of this sort happens and that all the possibilities are realized, each in a separate "universe" with a separate observer. Feynman's path integral, on the other hand, implies that there are lots of paths the universe could take representing all possible outcomes, each associated with an amplitude, and the measurement decides which one.

There is no collapse, no subjectivity. According to the many worlds, or many parallel universes, interpretation, at every quantum process—reflection versus transmission, decay now versus decay later—there is a kind of fragmentation of reality. All possibilities are realized, each in a different "universe," which then behaves as if the measurement had resulted in that possibility. Since this has presumably been occurring since the beginning of time, there is an extravagant multitude of universes. The scientist observer, when she enters the picture, must also fragment so that each quantum possibility is accompanied by an observer. The parallel universes and their accompanying observers are unaware of the fragmentation. This provides us with an infinite number of futures, many very similar to, but some wildly different from, the rest. If you find this idea hard to swallow, then you should see it is a measure of the desperation some physicists feel about the quantum world. Others simply say these are the components that comprise the path integral—compute and move on.

TO BE AND NOT TO BE

Here is how the many-worlds hypothesis explains the famous Schrödinger's cat in the box paradox: The universe splits into two universes, one with a live cat and one with a dead one. The latter does not die when somebody opens the box, collapsing the wave function, as in the Copenhagen interpretation; it dies logically when the vial breaks. Similarly, all the spooky-action-at-a-distance effects (nonlocalities) vanish, because there is no wave function collapse. This crazy idea has some virtues, or as one wise physicists has said, "It is cheap in assumptions but expensive in universes."[6] At a 1997 conference on quantum science, a poll found eight supporters of the many-worlds interpretation, thirteen supporters of Copenhagen, and eighteen undecided, out of forty-eight obviously confused experts.

Overall, we are looking at a common thing, quantum physics, from its many different perspectives. We are trying to come up with a "best" description, but perhaps such a thing doesn't exist. It need not—it is a human convention. There is no "best poem" or even a best interpretation of a great poem. We are the proverbial three blind men fondling the proverbial elephant. Pondering quantum reality is a bit like a dog chasing its tail—it doesn't penetrate any deeper into what we are operationally doing and how things actually work. For a person who approached physics with a deep and abiding philosophy of naturalness, as Einstein did, quantum reality is no less than an intellectual catastrophe. Physicists have brilliantly deciphered how quantum theory works, but they are only clerks taking dictation or workers on an assembly line when it comes to trying to understand why it is this way.

QUANTUM WEALTH

Some exciting new potential applications have originated from thinking deeply about the Einstein-Podolsky-Rosen paradox and Bell's theorem in the domain of "information theory." This field has drawn in a new group of thinkers with strong ties to computer science. Already these applications have attained respectability and promise to achieve even more major innovations in the next few decades. At least that's what the proponents believe.

Here is what author Andrew Steane has to say, in the introduction to his article "Quantum Computing":

> The subject of quantum computing brings together ideas from classical information theory, computer science and quantum physics. . . . Information can be identified as the most general thing which must propagate from a cause to an effect. It therefore has a fundamentally important role in the science of physics. However, the mathematical treatment of information, especially information processing, is quite recent, dating from the mid-twentieth century. This has meant that the full significance of information as a basic concept in physics is only now being discovered. This is especially true in quantum mechanics. The theory of quantum information and computing puts this significance on a firm footing, and has led to some profound and exciting new insights into the natural world. Among these are the use of quantum states to permit the secure transmission of classical information (quantum cryptography), the use of quantum entanglement to permit reliable transmission of quantum states (teleportation), the possibility of preserving quantum coherence in the presence of irreversible noise processes (quantum error correction), and the use of controlled quantum evolution for efficient computation (quantum computation). The common theme of all these insights is the use of quantum entanglement as a computational resource.[7]

Let us consider a few of these in detail. Be forewarned that we are visiting a complex and unusual reality shop, in which we try to make use of quantum weirdness.

QUANTUM CRYPTOGRAPHY

The problem of secure communication is an old one. Since ancient times, military intelligence has often involved codes and code breakers. In Elizabethan times, code breaking provided the crucial evidence that led to the execution of Mary Queen of Scots, while many people believe that the decisive event of World War II was the Allies' success in breaking the German's "unbreakable" Enigma code in 1942.[8] Part of the "game" the code senders play with the code breakers is try to determine if the code has been broken and if so to respond by sending "disinformation." On the flip side, the code breakers hope to determine the code but not let the senders know.

In our own time, as any newspaper reader knows, cryptography is no longer of interest primarily to spies and spymasters. The last time you gave your Visa card number to eBay or Amazon.com, you were trusting that this information was secure. But the derring-do of modern info-terrorists makes us all too aware that the secure exchange of information, from office e-mail to bank fund transfers, hangs by a fragile thread. Our government is concerned enough to spend billions of dollars on solving the problem.

The basic solution is to provide a cryptographic "key" to two separate locations that would permit the sending and reading of encoded messages between the locations. The standard way to encode confidential messages is to "hide them" in a long list sequence of otherwise random numbers. However, spies, hackers, and guys in black hats with larceny in their hearts and good knowledge of computers can crack the code by studying the sequences of random numbers, and often no one would be the wiser.

But quantum science can supply its own special randomness—a wonderful, wild randomness that can't be cracked. Furthermore, it can immediately betray any effort by code breakers to attempt to break the code! Because the history of cryptography is one of "unbreakable" codes, ultimately being cracked by superior technology, this assertion must be greeted with a fine dose of skepticism (the most famous instance being the German Enigma machine of World War II that produced what was considered to be an unbreakable code but which was heroically broken by the allies. The German's had no idea that the code had been broken).[9]

Let's take a closer look at cryptography. The coin of the realm of information science is the concept of a "bit," the smallest unit of information. A single bit is simply a binary digit, either a 1 or a 0. For example, the outcomes of the (classical) coin toss can be recorded in binary form, 1 or 0. The result of the coin toss, heads (tails) could be represented by 1 (0) and represents one bit of information. A series of coin tosses might read: 101100010111010010101011 1.

There is a quantum equivalent of the classical action that the experts have termed a "qubit." In quantum science an electron spin can be either up or down when measured in a given detector. "Up" or "down" replaces the 0 and 1 with quantum information attributes called qubits (there's no relation here to the biblical "cubits" with which Noah measured out his Ark, though they are pronounced the same). These qubits encode the quantum spin state: spin-up corresponds to 1, and spin-down corresponds to 0. So far we've introduced nothing new.

However, a qubit in quantum theory can exist either as a pure state or a mixed state. In a pure state, the measurement doesn't affect the state. For example, if we measure the spin of an electron along the "*z*-axis" with a certain detector, we will necessarily cause the electron to have its spin aligned along (up) or opposite to (down) the *z*-axis. For a random quantum electron we get either result with a certain probability. But if the electron is prearranged by a transmitter to be a pure spin state, either up or down, along the *z*-axis, then our detector observes the spin, but it doesn't change the electron's spin state.

So we can in principle send a message in binary code consisting of electrons (or photons) that are purely spin-up or spin-down along the *z*-axis. These are pure states, not mixed states, and anyone measuring them with a receiver, *also aligned along the* z-*axis*, will be able to read the code and will not affect the electron spin states. But what defines the *z*-axis? Only our specified (secret) choice does. We can choose any direction in space to be the *z*-axis. Then we send the information as to what defines the *z*-axis—the "key"—to our distant colleague to whom we intend to send the coded messages.

Now, however, anyone monitoring the signal with a detector that isn't perfectly aligned along the *z*-axis will receive only scrambled spin states of electrons (and will not know it), and they will also necessarily disturb the states of the electron spins upon making the measurements.

So not only will he obtain a message that is random gibberish but our colleague subsequently reading the message will be able to tell that it was "tampered with." In this way, we'll know if someone is listening and we can respond accordingly. Conversely, if our message is undisturbed, we will know that our message was completely secure. The main point here is that the spy's interdicting measurements introduce random changes that both the sender and the receiver can recognize. Once we learn there is a spy, we cancel the communication.

Transmitted quantum states can be used to establish a pair of identical sequences of random binary digits, which function as a cryptographic key to maintain secure communication. The quantum nature permits the assurance that the key is secure, for if it is compromised, the breach will be known. Quantum cryptography has been tested up to distances of many kilometers between source and detection. The practical application of quantum key distribution has not yet arrived, however, as it requires a huge investment in state-of-the-art laser equipment. But someday we may be able to eliminate those annoying charges to our credit cards for items we'd never buy, purchased in remote foreign countries.

QUANTUM COMPUTERS

There is a threat to the ultimate security provided by quantum cryptography, and it's called the quantum computer. In addition, the quantum computer is becoming a candidate for the ultimate supercomputer of the twenty-first century. Gordon Moore enunciated "Moore's law," stating that "the number of transistors on a chip doubles in twenty-four months."[10] Someone jested that if automotive technology progressed as rapidly as computers have over the past thirty years, a car would weigh sixty grams (about five ounces), cost $40, have about a cubic mile of luggage space, and use only a gallon of gasoline to travel one million miles in one hour!

Computer technology went from gears to relays to vacuum tubes to transistors to integrated circuits, and so on, in less than a human lifetime. Yet all computer science, as realized in the best computer available in the year 2000, is classical, obeying the laws of classical physics. Quantum computing is the conjecture that we may realize a new type of computing based on the laws of quantum science. Not yet on the drawing boards at IBM or even the most entrepreneurial startup in Silicon Valley—at least not to my knowledge—the quantum computer would make the fastest contemporary computer look like an abacus operated by a person without hands.

Quantum computing makes use of the aforementioned qubits, but an appreciation of it depends on a knowledge of information theory in the quantum world. Key ideas were proposed by Richard Feynman and others in the early 1980s and further developed by David Deutsch in 1985. These concepts and the subsequent contributions of a growing number of quantum computing groups resulted in quantum "gates" (switches that are either open or closed). These groups realized that quantum interference effects coupled with EPR-Bell correlations formed the ingredients of a potentially vastly more powerful way of carrying out certain computations.[11]

Interference effects, as illustrated by the double-slit experiment, involve the weirdest of quantum ideas. The presence of two open slits changes the places that a single photon can go on the catcher screen. We accept this by arguing that what is interfering are the quantum amplitudes that explore all possible paths, and we end up describing the net possibility (and associated probability) that one photon will end up at some particular location on the catcher screen. But if the intermediate screen had not two slits but a thousand slits, again there would be places the single photon could go and places it could not go. To obtain the probability of light reaching a particular point on the catcher screen, one would have to calculate paths from each of the thousand slits, add them all up, and square the result. This quantum complexity is increased further if there are two photons present simultaneously. Now each photon has a thousand possible paths for every one of its partner's choices of path. This works out to a million different states. If we have three photons, we get a billion states, and more and more and more as we keep adding photons. The calculation for predicting the outcome increases exponentially with the number of inputs.

The outcome of each of these problems may be very simple, certainly predictable, but highly impractical from the calculational point of view. Feynman's idea was to recognize the power of an *analogue* computer, that is, put in real photons and actually perform the experiment on a real quantum mechanical system, thus letting nature carry out this monstrous calculation simply and quickly. The ultimate quantum computer would choose which measurements must be made, and on what real system, and how to incorporate these measurement results into a subcalculation in the overall computation. This involves a slightly more wild version of the old double-slit experiment.

FUTURE WONDER COMPUTERS

To give you an idea of the power of quantum processes, let's do a simple comparison with classical computation. Suppose we have a classical three-bit "register," a device with three switches that can be either open (0) or closed (1). This means we can hold, at any one time, any one of eight possible numbers: 000, 001, 010, 011, 100, 101, 110, 111 or 1, 2, 3, 4, 5, 6, 7, 8. An ordinary computer encodes your numbers via open (0) and closed (1) switches. It is easy to see that a four-bit register (four switches) can encode sixteen numbers but only one at a time.

However, if the "register" is an atom instead of a mechanical or electronic switch, it is capable of being in, say, a superposition of the ground state (0) and the excited state (1)—in other words, a qubit. A three-qubit register is capable of expressing all eight possibilities simultaneously, since each qubit can be in both the 0 and 1 states. A four-qubit register can have sixteen numbers, and, in "mathematical language," an "N-qubit" register can have 2^N numbers.

In a classical computer, the electronic bit will typically be the electric charge on a tiny capacitor, registering as charged (1) or no charge (0). By regulating the flow of electrons to or from the capacitor, we can manipulate the numbers. In the quantum system, in contrast, pulses of light can excite or de-excite the atom. In contrast to a standard computer, a quantum calculation allows both 0 and 1 to take part in the same step of the calculation at the same time. One can begin to see the possibilities.

If we have a ten-qubit register, we can represent all the numbers from 0 to 1024 simultaneously. With two such registers, we might couple them in such a way (multiplication) that the output would contain all the numbers in the 1024 times 1024 multiplication table. While a high-speed conventional computer would have to carry out over a million individual calculations to work out all the numbers in such a table, the quantum computer would explore all the possibilities simultaneously and reach the same result in a single effortless step.

Such theoretical considerations have led to the belief that, for certain kinds of computations, a quantum computer could solve in a year a problem that would take today's best computer several billion years! The quantum computer's power comes from its ability to act on all its possible states simultaneously, performing many operations in parallel and using only a single processing unit. However (to the music of "Also Sprach Zarathustra" by Richard Strauss), before you invest your life savings in a quantum computing startup company in Cupertino, California, you should know that some of the quantum computing experts are skeptical about its ultimate utility (though they do believe that the theoretical excursion is valuable in throwing light on the fundamentals of quantum theory).

Even if awesome problems can be solved, these pundits point out, we're still dealing with a very different kind of computer that is specific to particular sets of problems and thus is unlikely to replace the classical computers in use today. The classical world is different from the quantum world, which is why we don't take our broken Chevy to a quantum mechanic. Among the perceived obstacles to quantum computing is its sensitivity to interference from outside noise (if one qubit changes its state due to an interaction with a cosmic ray from space, the whole calculation is destroyed). Furthermore, it is basically an analogue device, meaning it is designed to imitate the particular process it is computing, and therefore it lacks the full generality of a computer that runs a program for anything we choose to compute. It also happens to be true that it is not easy to make such a computer. To make quantum computing a reality will require solving stringent reliability problems and finding useful quantum algorithms, that is, solving problems that would actually make quantum computers worthwhile.

The ability to factor huge numbers into their primes (for example, 21 = 3 times 7) is potentially one such algorithm. While it is (classically) easy to check that two numbers are factors of a third number—for example, that 5 and 13 are factors of 65—it is generally very difficult to find the factors of a very large number such as

3,204,637,196,245,567,128,917,346,493,902,297,904,681,379

In addition to having cryptography applications, this problem may ultimately prove to be a good showcase for the power of a quantum computer, since it is not solvable with conventional classical computers.

We mention in passing a bizarre hypothesis of English mathematician and theorist Roger Penrose about human consciousness. A human being can make almost lightning-fast calculations that rival what a computer can do, but he doesn't operate the way a computer does. Even in playing chess with a computer, a human is assessing a vast array of sensory effects associated with experience and integrating them quickly to beat an algorithmic approach programmed into a much faster electronic computer. While the computer's result is exact, the human one is efficient but fuzzy—not always exactly correct. Precision and accuracy are sacrificed to some optimal degree for speed.

Penrose suggests that perhaps the overall sensation of conscious awareness is a coherent sum of many possibilities—a quantum phenomenon.[12] All of this indicates to Penrose that *we are* quantum computers and that the wave functions we are using to store and interfere to produce computational results may be distributed beyond our brains, over our bodies. Penrose suggests in his book *Shadows of the Mind* that the wave functions of human consciousness reside in the mysterious microtube system within the nerve cells. Well, that's interesting, but we will still need a theory of consciousness.

Nevertheless, quantum computing may ultimately redeem itself by illuminating the role of information theory in fundamental quantum physics. Perhaps we will end up with a powerful new type of computing and a view of the quantum world—perhaps one that is more resonant with our evolving intuitions—less bizarre, less spooky, less weird. If this happens, it will be one of the very rare times in the history of science that an independent discipline (information science or a theory of consciousness) merged with physics to contribute—perhaps profoundly—to its fundamental structure.

FINALE

We thus conclude our tale with many philosophical questions left hanging: How can light be both a wave and a particle? Are there many worlds or only one? Is there an unbreakable code? What is ultimate reality? Are the laws of physics themselves determined by a throw of the dice? Or are these questions empty? Is the answer "Quantum physics simply requires getting used to it"? And you may ask: "Where and when will the next great scientific leap take place?"

Our trajectory began with Galileo's death blow to Aristotelian physics at Pisa. We moved into the clocklike regularity of Isaac Newton's classical universe, with its predictable forces and laws. The understanding of our world, and our place in it, might have rested there, in considerable comfort (albeit without cell phones), but it didn't. Then onto the mysterious forces that haunted the mid-nineteenth century—electricity and magnetism—that were deciphered and woven into a classical fabric by Michael Faraday and James Clerk Maxwell. Now our physical world seemed complete. By that century's end, there were some people who predicted the end of physics. All the outstanding puzzles worth solving had seemingly been solved, and the mere details were surely forthcoming, all within the classical order of Newton. We had come to the end of the line. The physicists could fold up their tents and go home.

Of course, it turned out that there were still a few incomprehensible things lying around that needed an explanation. Campfire coals glow red when calculations suggest they should glow blue, and why could we not detect the motion of Earth through the ether and catch up with a light beam? Our understanding of the universe, as we knew it, might not be the last word on the subject. The whole universe was about to be refigured by a new and stellar team—Einstein, Bohr, Schrödinger, Heisenberg, Pauli, Dirac, and others eager to do the job.

Of course, good old Newtonian mechanics continued to work just fine for most things, like planets, rockets, bowling balls, steam locomotives, and bridges. Even in the twenty-seventh century a home run will still soar in a lovely parabolic arc à la Newton. But, after 1900, or 1920, or 1930, if one wanted to know how the atomic and subatomic world worked, one was forced to grow a new brain—one capable of grappling with quantum physics and its inherently probabilistic nature. Recall that

Einstein never accepted a universe that, at its core, was fundamentally based on probability.

We know this trek hasn't been an easy journey. If the ever-recurring double-slit paradox weren't enough of a headache, there was the vertiginous terrain of the Schrödinger wave function, the Heisenberg uncertainty relations, and the Copenhagen interpretation, among other daunting theories. Cats were simultaneously pronounced alive and dead; light was a particle and a wave; a system could not be detached from the observer; there were arguments over whether God plays dice with the universe. Then, just when you had gotten a grip on all that, things got even messier: the Pauli exclusion principle, Einstein-Podolsky-Rosen, and Bell's theorem. These matters are not light cocktail party conversation, even for those who subscribe to New Age credos and often get the facts scrambled. Still, we soldiered on and didn't give up, even struggling through a few unavoidable equations.

We have been adventurous and willing to entertain theories so far-out they might be the titles of *Star Trek* episodes—"many worlds," "Copenhagen" (after a hit play by the same name), "Strings and M-Theory," "the Landscape," and so on. We hope the journey has been worthwhile and that you now share with us physicists a sense of the grandeur and profound mystery of our world.

We still have looming ahead in our new century the problem of understanding human consciousness itself. Perhaps human consciousness may be explained as a quantum state phenomenon. However, just because we don't completely understand either of these phenomena doesn't mean they are connected—but that's what many scientists believe.

The mind does appear in quantum science, you may remember, when we make measurements. The observer (a mind) always interferes with the system being observed, and one may well ask whether this requires understanding of how human consciousness fits into the physical world. Does the mind-body problem connect with quantum science? For all our recent progress in understanding how the brain encodes, processes information, and controls behavior, a deep mystery remains. How do these physicochemical activities develop an "inner" or "subjective" life; how do they generate what it is like to be "you"?

Nevertheless, the quantum-mind connection has its critics as well, including Francis Crick, codiscoverer of DNA, who writes in *The Astonishing Hypothesis*: "'You,' your joys and your sorrows, your memories and your ambitions, your sense of personal identity and free will, are in fact no more than the behavior of a vast assembly of nerve cells and their associated molecules."[13]

Thus, it is our hope that this is only the beginning of your journey and that you will explore further the marvels and seeming paradoxes that make up our quantum universe.

APPENDIX: SPIN

We include as part of this book an appendix on spin. We have also prepared additional appendixes on general topics in classical and quantum physics and on Bell's theorem. These will be available through our website, http://www.emmynoether.com, as downloadable pdf files.

WHAT IS SPIN?

Welcome to the quintessential quantum property of "spin." Any rotating body has spin—a top, a CD player, Earth, the washing machine basin on the rinse cycle, a star, a black hole, a galaxy—all have spin. So, too, quantum particles, molecules, atoms, nuclei of atoms, the protons and neutrons in the nucleus, the particle of light (photons) electrons, the particles inside protons and neutrons (quarks, gluons), and so on. But while large classical objects can have any amount of spin, and can stop spinning altogether, quantum objects have "intrinsic spin" and are always spinning with the same total intrinsic spin.

An elementary particle's spin is one of its defining properties. For example, an electron is an elementary particle that has spin. We can never halt an electron from spinning, or else it would no longer be an electron. However, we can rotate a particle in space, and the value of its spin, as projected along any given axis in space, will change, just as it does for a classical spinning top. The difference in quantum physics is that we can only ask what the value of the spin is when projected along a given axis because that is what we can measure—asking about things we cannot measure is meaningless in quantum physics.

Let's discuss the rotational motion of a classical object. Linear physical motion is measured by something called momentum, which is simply mass times velocity. Note that this combines the concept of matter (mass) and concept of motion (velocity), so it represents a kind of overall measure of "physical motion." This is a *vector* quantity, since the velocity is a vector, having both a magnitude (speed) and a direction (of motion) in space. In general, a vector can be visualized as an arrow in space with both magnitude and direction.

Likewise, physical rotational motion is measured by a vector quantity called "angular momentum." Classically, angular momentum involves the way in which mass is distributed throughout the object, which is the "moment of inertia." If a body is large, with a large radius, when it spins there is a lot more matter spinning than if the same amount of mass were distributed within a smaller radius. So, not surprisingly, the moment of inertia, I, increases with the size of the body. In fact, it is mass times "the (approximate) radius of the body squared," or roughly $I = MR^2$, with M the mass and R the "radius" of the body. This can be made very precise using calculus.[1]

Spin also involves the *angular velocity*—how fast the object is actually rotating. Angular velocity is usually denoted by ω (omega) and is "so and so many radians of rotation per second." (360 degrees equals 2π *radians*; so, for example, 90 degrees corresponds to $\pi/2$ radians. Radians are a more mathematically natural way to measure angles than degrees because a circle with a radius of one has a circumference of 2π.) So spin is just the product of the moment of inertia times the angular velocity, or $S = I\omega$. (Compare: momentum is mass times velocity and describes motion in a straight line, while spin is moment of inertia times angular velocity—these are very similar constructs.) Spin is also a vector quantity, pointing along the axis of the spin rotation. Here we use the "right-hand rule" to establish the direction of the spin vector: curl the fingers of your right hand in the direction of the spinning motion, and your thumb will point in the direction of the spin vector.

Spin (or, more generally, angular momentum) is a *conserved quantity* (like energy and momentum) such that the total angular momentum of an undisturbed isolated system remains forever constant. As a consequence of this, we see that an ice skater, viewed as a physical system, can dramatically increase her spin motion (angular velocity) as she draws her arms inward. The spin angular momentum is $S = I\omega = MR^2\omega$, and it must stay the same as she pulls her arms in. Pulling her arms in decreases R, while M stays the same. So the angular velocity ω must increase to compensate the decrease in R. In fact, R^2 becomes four times smaller if the skater simply decreases her arms' outward distance, *R*, by a half, so her angular velocity must increase approximately fourfold, which is why this is such a dramatic stunt. Angular momentum is a very important effect in nature. Frisbies are a popular application of the principle of the conservation of angular momentum. Pilots, however, must always avoid the dreaded "flat-spin," where they can inadvertently get an airplane spinning like a Frisbie, and the conservation of angular momentum makes it very difficult to recover control of the airplane.

FIGURE 37: The "right-hand rule" defines the direction of a spin vector. Curl the fingers of your right hand in the direction of rotation, and your thumb defines the orientation of the spin vector. For classical objects, the spin can be arbitrary with any value along any direction. The electron when its spin is measured along any chosen direction will always have either spin 1/2 or spin-1/2 in units of $\hbar = h/2\pi$.

Angular momentum, which was a continuously varying quantity in Newtonian physics, also changes its character drastically in quantum mechanics. It becomes quantized. *Angular momentum is always quantized in quantum mechanics*. All observed angular momenta as measured along any *spin axis* are discrete multiples of $\hbar = h/2\pi$, where h is Planck's constant. All the particle spin and orbital states of motion we find in nature have angular momenta that can have only the exact values

$$0, \frac{\hbar}{2}, \hbar, \frac{3\hbar}{2}, 2\hbar, \frac{5\hbar}{2}, 3\hbar$$

and so forth. Angular momentum is always either an *integer* or a *half-integer* multiple of $\hbar$ in nature. We don't see this quantization effect for very large classical objects because they can have such enormous angular momenta, many times greater than $\hbar$. Only at the level of exceedingly tiny systems, atoms, or the elementary particles themselves, do we observe the quantization of angular momentum.

Angular momentum is therefore an intrinsic property of an elementary particle or an atom. All elementary particles have spin angular momentum. We can never slow down an electron's rotation and make it stop spinning. An electron always has a definite value of its spin angular momentum, and that turns out to be, in magnitude, exactly $\hbar/2$. We can flip an electron and then find its angular momentum is pointing in the opposite direction, or $-\hbar/2$. These are the only two observed values of the electron's spin when measured along any chosen direction in space. We say that "the electron is a spin-1/2 particle," because its angular momentum is the particular quantity $\hbar/2$.

Particles that have *half-integer multiples* of $\hbar$ for their angular momentum, that is

$$\frac{\hbar}{2}, \frac{3\hbar}{2}, \frac{5\hbar}{2}$$

and so on, are called *fermions*, after Enrico Fermi, who helped pioneer these concepts (with Wolfgang Pauli and Paul Dirac). The main fermions we encounter in most of our discussions are the electron, the proton, or the neutron (and quarks, which make up the proton and neutron, etc.), and each has angular momentum $\hbar/2$. We refer to all of these as "spin-1/2 fermions."

Particles, on the other hand, that have angular momenta that are *integer multiples* of $\hbar$, such as

$$0, \hbar, 2\hbar, 3\hbar$$

and so on, are called *bosons*, after the famous Indian physicist Satyendra Nath Bose, who was a friend of Albert Einstein, and also who developed some of these ideas. There is a profound difference between fermions and bosons that we'll encounter momentarily. Typically, the only particles that are bosons and that will concern us presently are particles like the photon, which has "spin-1," or one unit of angular momentum; the particle of gravity, the graviton, which has yet to be detected in the lab and has "spin-2," or $2\hbar$ units of angular momentum; and other particles made of quarks and antiquarks, called mesons, that have "spin-0," or 0 units of angular momentum. Orbital motion also has angular momentum. All orbital motion, in quantum theory, has integer units of $\hbar$ for angular momentum, hence 0, $\hbar$, $2\hbar$, $3\hbar$, and so on.

EXCHANGE SYMMETRY

A symmetry that is of paramount importance in shaping the physical world is the *symmetry of identical particles in quantum mechanics*. All elementary particles are so fundamental that they have no identifying labels, and any two of them absolutely cannot be distinguished from each other. There is no difference between any two electrons in the universe. The same is true of photons, muons, neutrinos, quarks, and so on. The quantum effect of identity depends strongly upon spin.

We can understand the origin of these effects as a symmetry in the language of Schrödinger's wave function. Let us consider a physical system containing two particles. For example, this could be a helium atom, which has two orbiting electrons. In general, we describe the two-particle system by a quantum mechanical wave function that depends now upon the two different positions of the two identical particles as

$$\psi(\vec{x}_1, \vec{x}_2, t)$$

Again, according to Max Born the (absolute) square of the wave function is the probability $|\psi(\vec{x}_1, \vec{x}_2, t)|^2$ = probability of finding particle (1) at $\vec{x}_1$ and particle (2) at $\vec{x}_2$ at time t.

Now consider the act of *exchanging one particle with the other particle.* In other words, we rearrange our system with the swapping of the two positions $\vec{x}_1 \leftrightarrow \vec{x}_2$ Hence the new "swapped" system is described by the wave function: $\Psi(\vec{x}_2, \vec{x}_1, t)$, where we have simply interchanged the two particles' positions. But is this really a new system or just the original system we started with? To put it another way, is this the wave function describing a new swapped system or is it the same original system?

Now, in everyday life, the category of "things" that we encounter called "dogs" is very large, and no two dogs are identical. However, all electrons are precisely identical to each other. Electrons carry only a very limited amount of information. Any given electron is *exactly* identical to any other electron. The same is true of the other elementary particles. Therefore, any physical system must be symmetrical, or invariant, under the swapping of one such particle with another. *Swapping identical particles in the wave function is a fundamental symmetry of nature.* In a sense, nature is very simple-minded in the way it treats electrons in that it cannot detect the difference between any two (or more) electrons in the whole universe.

This "exchange symmetry" of the wave function *must leave the laws of physics invariant because the particles are identical.* At the quantum level, this implies that our swapped wave function must give the same observable probability as the original one: $|\Psi(\vec{x}_1, \vec{x}_2, t)|^2 = |\Psi(\vec{x}_2, \vec{x}_1, t)|^2$. This condition, however, implies two possible solutions for the effect of the exchange on the wave function:

either: or $\Psi(\vec{x}_1, \vec{x}_2, t) = \Psi(\vec{x}_2, \vec{x}_1, t)$ or $\Psi(\vec{x}_1, \vec{x}_2, t) = -\Psi(\vec{x}_2, \vec{x}_1, t)$

So the exchanged wave function can either be *symmetrical*, that is +1, times the original one, or else it can be *anti-symmetrical*, or –1 times the original one. Either case is allowed, in principle, because we can measure only the probabilities (the squares of wave functions).

In fact, quantum mechanics allows both possibilities, so nature finds a way to offer both possibilities, and the result is astonishing.

BOSONS

It turns out that when we are talking about *bosons*, then the rule is that, upon swapping two particles in the wave function, we would get the + sign:

$$\text{Exchange Symmetry of Identical Bosons: } \Psi(\vec{x}_1, \vec{x}_2, t) = \Psi(\vec{x}_2, \vec{x}_1, t)$$

With this result, we immediately see an important effect—two identical bosons can easily be located at the same point in space, that is, $\vec{x}_1 = \vec{x}_2$. This means that $\Psi(\vec{x}_1, \vec{x}_2, t)$ can be nonzero for someplace in space. In fact, by considering lots of bosons localized in the same region of space, described by one big wave function, we can actually prove that the *most probable place for all the bosons in a system is piled on top of one another!* So it is possible to coax a lot of identical bosons to share the same little region in space, almost an exact pinpoint in space. Or, the identical bosons can be coaxed readily into a quantum state with all the particles having the exact same value of momentum. Thus, we say that bosons *condense* into compact, or "coherent," states. This is called *Bose-Einstein condensation.*

There are many variations on Bose-Einstein condensation and all kinds of phenomena that have many bosons in one quantum state of motion. *Lasers* produce coherent states of many, many photons all piled into the same state of momentum, moving together in exactly the same state of momentum at the same time. *Superconductors* involve pairs of electrons bound by crystal vibrations (quantum sound) into spin = 0 bosonic particles (called "Cooper pairs"). In a superconductor the electric current involves a coherent motion of many of these bound pairs of electrons sharing exactly the same state of momentum. *Superfluids* are quantum states of extremely low-temperature bosons (as in liquid ^{4}He) in which the entire liquid condenses into a common state of motion which becomes completely frictionless. It has to be the isotope ^{4}He in order to have a superfluid (2 protons + 2 neutrons in the nucleus), because the isotope ^{4}He is a boson, while the other common isotope ^{3}He is not (with 2 protons + 1 neutron in the nucleus, it is a fermion; see below). Bose-Einstein condensates can occur in which many bosonic atoms condense into ultracompact droplets of very large density, with the particles piling on top of one another in space.

FERMIONS

If, on the other hand, we exchange a pair of identical electrons *(fermions)* in a quantum state, the rule is that we get the (–) sign in front of the wave function. This holds for any particle with fractional spin, such as the electron with spin-1/2.

$$\text{Exchange Symmetry of Identical Fermions: } \Psi(\vec{x}_1,\vec{x}_2,t) = -\Psi(\vec{x}_2,\vec{x}_1,t),$$

We can therefore see a simple yet profound fact about identical fermions: No two identical fermions (with their spins all "aligned") can occupy the same point in space: $\Psi(\vec{x}_1,\vec{x}_2,t)$. This follows from the fact that if we now swap the position with itself, we must get $\Psi(\vec{x}_1,\vec{x}_2,t) = -\Psi(\vec{x}_2,\vec{x}_1,t)$, and therefore $\Psi(\vec{x}_1,\vec{x}_2,t) = 0$, because only 0 equals minus itself.

More generally, no two identical fermions can occupy the same quantum state of momentum either. This is known as the "Pauli exclusion principle," after the brilliant Austrian-Swiss theorist Wolfgang Pauli. Pauli proved that his exclusion principle for spin-1/2 comes from the basic rotational symmetries of the laws of physics. It involves the mathematical details of what spin-1/2 particles do when they are rotated. Swapping two identical particles in a quantum state is identical to rotating the system by 180° in certain configurations, and the behavior of the spin-1/2 wave function then gives the minus sign.[2]

The exclusion property of fermions largely accounts for the stability of matter. For spin-1/2 particles, there are two allowed states of spin, which we call "up" and "down" ("up" and "down" refer to any arbitrary direction in space). Thus, in an atom of helium, we can get two electrons into the same lowest-energy orbital state of motion. To get the two electrons in one orbital requires that one electron has its spin pointing "up" and the other has spin pointing "down." However, we *cannot then insert a third electron* into that same orbital state because its spin would be either up or down, the same as one of the two electrons already present. The exchange symmetry minus sign would force the wave function to be zero.

In other words, if we try to exchange the two electrons whose spins are the same, the wave function would have to equal minus itself and must therefore be zero! Hence, to make the next atom, lithium, the third electron must go into a new state of orbital motion, that is, a new *orbital.* Thus, lithium has a *closed inner orbital or "closed shell"* (i.e., a helium state inside it) and a sole outer electron. This outer electron behaves much like the sole electron in hydrogen. Therefore, *lithium and hydrogen have similar chemical properties.* We thus see the emergence of the periodic table of elements as in chapter 6. If electrons were not fermions, and did not behave this way, every electron in the atom would rapidly collapse into the ground state. All atoms would behave like hydrogen gas. The delicate chemistry of organic (carbon containing) molecules would never happen.

Yet another extreme example of fermionic behavior is that of the neutron star. A neutron star is formed as the core of a giant supernova imploding while the rest of the star is blown out into space. The neutron star is made entirely of gravitationally bound neutrons. Neutrons are fermions, with spin-1/2, and again the exclusion principle applies. The state of the star is supported against gravitational collapse by the fact that it is impossible to get more than two neutrons (each with spins counter-aligned) into the same state of motion. If we try to compress the star, the neutrons begin to increase their energies because they cannot condense into a common lower-energy state. Hence, there is a kind of pressure, or resistance to collapse, driven by the fact that fermions are not allowed into the same quantum state.

All these bizarre macroscopic phenomena come from the *exchange symmetry* of the quantum wave functions of elementary particles. We don't observe this exchange symmetry in the case of poodles, or people, or any other everyday macroscopic objects. This is "simply" a consequence of their complexity. Complexity requires that the individual particles have to be far apart from one another so that many different physical states are possible, and the particles never come at all close to being in the same quantum states at the same time. One poodle differs from another because of this complex arrangement of its quantum components. Thus the effects of identity are not obvious in complex extended systems that are far removed from the quantum ground state.

Notes

CHAPTER 1

1. For more discussion of the ascent in the conceptual framework of physics, see Leon M. Lederman and Christopher T. Hill, *Symmetry and the Beautiful Universe* (Amherst, NY: Prometheus Books, 2004).

2. Ibid.

3. See Max Born, *The Born-Einstein Letters: Friendship, Politics and Physics in Uncertain Times* (New York: Macmillan, 2004); Barbara L. Cline, *Men Who Made a New Physics: Physicists and the Quantum Theory* (Chicago: University of Chicago Press, 1987); A. Fine, "Einstein's Interpretations of the Quantum Theory," in *Einstein in Context: A Special Issue of Science in Context*, edited by Mara Beller, Robert S. Cohen, and Jürgen Renn (Cambridge: Cambridge University Press, 1993), pp. 257–73.

4. Walter J. Moore, *A Life of Erwin Schrödinger*, abridged ed. (Cambridge: Cambridge University Press, 1994).

5. C. P. Enz, "Heisenberg's Applications of Quantum Mechanics (1926–33) or the Settling of the New Land," *Helvetica Physica Acta* 56, no. 5 (1983): 993–1001; Louisa Gilder, *The Age of Entanglement: When Quantum Physics Was Reborn* (New York: Alfred A. Knopf, 2008). It is interesting to peruse the *Werner Heisenberg Austolang: Helgoland* website at http://www.archiv.uni-leipzig.de/heisenberg/Geburt_der_modernen_Atomphysik (accessed January 1, 2010). Jeremy Bernstein, *Quantum Leaps* (Cambridge, MA: Belknap Press of Harvard University Press, 2009); Arthur I. Miller, ed., *Sixty-Two Years of Uncertainty: Historical, Philosophical, and Physical Inquiries into the Foundations of Quantum Mechanics* (New York: Plenum Press, 1990); J. Hendry, "The Development of Attitudes to the Wave-Particle Duality of Light and Quantum Theory, 1900–1920," *Annals of Science* 37, no. 1 (1980): 59–79.

6. To get a deeper appreciation of the nature of a "quantum state" description of a physical system in contrast to a classical description, see chapter 7, note 8: "Mixed States in Quantum Theory."

7. See Born, *The Born-Einstein Letters*; Cline, *Men Who Made a New Physics*; Fine, "Einstein's Interpretations of the Quantum Theory."

8. Schrödinger constructed the equations describing a wave, which are otherwise standard wave equations but adapted especially to answer this particular question. These concepts are known nowadays as "Schrödinger's (wave) equation" or "Schrödinger's formalism." The thing that is the wave, the thing that does the "waving," is called Schrödinger's "wave function." The whole framework invented by Schrödinger is mathematically equivalent to Heisenberg's, but that wasn't immediately obvious. See P. A. Hanle, "Erwin Schrödinger's Reaction to Louis de Broglie's Thesis on the Quantum Theory," *Isis* 68, no. 244 (1077): 606–609; Walter John Moore, *Schrödinger: Life and Thought* (Cambridge: Cambridge University Press, 1992); Walter J. Moore, *A Life of Erwin Schrödinger*, Canto ed. (Cambridge: Cambridge University Press, 2003).

9. The wave function, $\Psi(x, t)$ ("sigh"), is generally *a complex number* at any point in space and time. We will ignore this fact in the main text. As a consequence, we write for the mathematical square of the wave function, Ψ^2. Technically, we really should write the "absolute square," $|\Psi|^2$ appropriate to complex numbers. See discussion in chapter 5, note 15, "A Digression on Complex Numbers."

10. Or, more specifically, $\Psi(x, t)$ is the "square root" of a probability. $\Psi(x, t)$ is a complex number valued function of space and time. The square is therefore the absolute square, $\Psi \times \Psi = |\Psi|^2$, which is a positive real number at each point in space and time (see the "digression on numbers" in chapter 5, note 15). Dirac showed how the Heisenberg and Schrödinger formulations are mathematically equivalent, but "equivalence" and "convenience of use" are two different things. So, what does a quantum particle wave function look like? Using Schrödinger's wave equation, we would find that a freely traveling particle takes the form of a "traveling wave," having a wave function that looks like this:

$$\Psi(x,t) = A\cos(\vec{k}\cdot\vec{x} - \omega t) + iA\,\mathrm{si}(\vec{k}\cdot\vec{x} - \omega t) \text{ where } |\vec{k}| = 2\pi/\lambda \text{ and } \omega = 2\pi f$$

11. M. Paty, "The Nature of Einstein's Objections to the Copenhagen Interpretation of Quantum Mechanics," *Foundations of Physics* 25, no. 1 (1995): 183–204; K. Popper, "A Critical Note on the Greatest Days of Quantum Theory," *Foundations of Physics* 12, no. 10 (1982): 971–76. F. Rohrlich, "Schrödinger and the Interpretation of Quantum Mechanics," *Foundations of Physics* 17, no. 12 (1987): 1205–20. F. Rohrlich, "Schrödinger's Criticism of Quantum Mechanics—Fifty Years Later," in *Symposium on the Foundations of Modern Physics: 50 Years of the Einstein-Podolsky-Rosen Gedankenexperiment, Joensuu, Finland, 16–20 June 1985*, edited by Pekka Lahti and Peter Mittelstaedt (Singapore; Philadelphia: World Scientific, 1985), pp. 555–72. D. Wick, *The*

Infamous Boundary: Seven Decades of Controversy in Quantum Physics (Boston: Birkhauser, 1995).

12. Ibid.

13. More generally we can have the "mixed state":

$$a\,(J \rightarrow \text{Peoria},\ M \rightarrow \alpha\,\text{Centauri}) + b\,(J \rightarrow \alpha\ \text{Centauri},\ M \rightarrow \text{Peoria})$$

where $|a|^2 + |b|^2 = 1$. With equal probabilities for each, we have $|a|^2 = |b|^2 = 1/2$. See chapter 7, note 8 below for further elaboration. (The term "mixed state" is usually reserved for a different kind of representation in quantum theory called a "nondiagonal density matrix." We'll use the term presently to describe "mixtures of eigenstates" as in the above example.)

14. See Karen Michelle Barad, *Meeting the Universe Halfway, Quantum Physics and the Entanglement of Matter and Meaning* (Durham, NC: Duke University Press, 2007), p. 254; see footnote citing Niels Bohr, *The Philosophical Writings of Niels Bohr* (Woodbridge, CT: Ox Bow Press, 1998).

15. See note 13.

16. Robert Frost, "The Lockless Door," in *A Miscellany of American Poetry, Aiken, Frost, Fletcher, Lindsay, Lowell, Oppenheim, Robinson, Sandburg, Teasdale and Untermeyer (1920)* (New York: Robert Frost, Kessinger Publishing, 1920).

17. See note 1.

CHAPTER 2

1. Frictionless motion and the concept of an ideal vacuum was too great a conceptual leap at the time of the Greeks. Heavy objects certainly do not appear to execute "uniform motion in a straight line" *unless* one applies a force to them. All things in motion tend eventually to a natural state of rest, concluded Aristotle. Mass seemed, for the most part, a measure of a thing's tendency to return toward a state of rest and to produce grunting and groaning when lifted, pushed, or pulled. The Greeks lived in a world dominated by friction. They could not separate the concept of friction from the concept of idealized frictionless motion.

2. We discuss the history of the classical scientific revolution in Leon M. Lederman and Christopher T. Hill, *Symmetry and the Beautiful Universe* (Amherst, NY: Prometheus Books, 2004).

3. This is poem #1627 in Thomas H. Johnson, ed., *The Complete Poems of Emily Dickinson*, paperback ed. (Boston: Back Bay Books, 1976).

4. Edgar Allan Poe, *Complete Stories of Edgar Allan Poe*, Doubleday Book Club ed. (New York: Doubleday, 1984).

CHAPTER 3

1. There is a lot of good educational information on light to be found on the Internet, including graphics and animations. For example, see *Science of Light Animations*, http://www.ltscotland.org.uk/5to14/resources/science/light/index.asp; *How Stuff Works*, http://science.howstuffworks.com/light.htm; *Itchy-animation*, http://www.itchy-animation.co.uk/light.htm; *Thinkquest*, http://library.thinkquest.org/28160/english/index.html (all accessed May 22, 2010), and browse for more.

2. Laurence Bobis and James Lequeux, "Cassini, Röme, and the Velocity of Light," *Journal of Astronomical History and Heritage* 11, no. 2 (2008): 97–105.

3. Alex Wood and Frank Oldham, *Thomas Young* (Cambridge: Cambridge University Press, 1954); Andrew Robinson, "Thomas Young: The Man Who Knew Everything," *History Today* 56, nos. 53–57 (2006); Andrew Robinson, *The Last Man Who Knew Everything: Thomas Young, the Anonymous Polymath Who Proved Newton Wrong, Explained How We See, Cured the Sick and Deciphered the Rosetta Stone* (New York: Pi Press, 2005).

4. A traveling wave is sometimes called a *wave train*, with many sequential crests and troughs of the train as it traverses space. Such a wave is described by three quantities: its *frequency*, its *wavelength*, and its *amplitude*. The wavelength is the distance between two neighboring troughs or crests of the wave. The frequency is the number of times per second that the wave undulates up and down through complete cycles at any fixed point in space.

If we think of the wave as a long freight train, then its wavelength is the length of a boxcar and its frequency is the number of boxcars per second passing in front of us as we patiently wait for the train to pass. The speed of the traveling wave is therefore the length of a boxcar divided by the time it takes to pass, or (speed of wave) = (wavelength) times (frequency). Thus, knowing the speed, the wavelength and frequency are *inversely related*, or (wavelength) = (speed of wave) divided by (frequency), and (frequency) = (speed of wave) divided by (wavelength).

The *amplitude* of the wave is the height of the crests, or the depth of the troughs, measured from the average. That is, the distance from the top of a crest to the bottom of a trough is twice the amplitude of the wave, and it can be thought of as the height of the boxcars. For an electromagnetic wave, the amplitude is the strength of the electric field in the wave. For a water wave, twice the amplitude is the distance that a boat is lifted from the trough to the crest as the wave passes by. (See figure 5.)

The *color* of a visible light wave was understood in the nineteenth-century Maxwellian theory of electromagnetism to be determined by the wavelength (and inversely, the frequency). If we take the frequency to be small, we correspondingly find that the wavelength becomes large. Longer-wavelength visible light is red, while shorter-wavelength visible light is blue. *Visible red light* has a wavelength of about $0.000065 = 6.5 \times 10^{-5}$ centimeters (or 650 nanometers, or 6500 Å; one nanometer, nm, is 10^{-9} meters or 10^{-7} centimeters; one angstrom "Å" measures 10^{-8} cm). The longer the wavelength, the more deep red the color becomes, until it fades from our eyes' sensitivity at a wavelength of about $0.00007 = 7 \times 10^{-5}$ centimeters (700 nm, or 7000 Å). Taking the wavelength still larger, we have *infrared light*, which we can feel as gentle heat but cannot see with our eyes. As the wavelength becomes still larger, we enter the realm of microwaves, and, still longer, we have radio waves. On the other hand, as we take the wavelength to be shorter than about $0.000045 = 4.5 \times 10^{-5}$ centimeters (450 nm, or 4500 Å), light becomes blue. At shorter wavelengths (higher frequencies) it becomes deep violet blue, and then with still shorter wavelengths, it fades from visibility, at about $0.00004 = 4 \times 10^{-5}$ centimeters (400 nm or 4000 Å). At still shorter wavelengths, light becomes *ultraviolet*, eventually becoming x-rays and eventually gamma rays at much shorter wavelengths. (See figure 12.)

5. The Draupner Oil Platform Wave was hit by a single gigantic rogue wave, observed and measured on New Year's Day 1995, finally confirming the existence of these freak waves, which had previously been considered delusions of seamen. See *Physics, Spotlighting Exceptional Research*, http://physics.aps.org/articles/v2/86; and a fascinating discussion, *Freak Waves, Rogue Waves, Extreme Waves, and Ocean Wave Climate*, http://folk.uio.no/karstent/waves/index_en.html; references therein (accessed May 21, 2010).

6. There are excellent online animations and graphics showing wave interference. See in particular, Daniel A. Russel, *Acoustics and Vibration Animations*, http://paws.kettering.edu/~drussell/Demos.html, for example, Superposition of Two Waves, http://paws.kettering.edu/~drussell/Demos/superposition/superposition.html. Also see Physics in Context, http://www.learningincontext.com/PiC-Web/chapt08.htm (accessed May 21, 2010).

7. We optimally want a single color of light, such as produced by a laser pointer, since the interference pattern will form at different locations if we have different colors and would be harder to observe. Young probably used a candle, but he could have improved his experiment with color filters. Incidentally, there is diffraction at a single slit due to the finite size of the slit, but it is hard to resolve if the slits are very narrow compared to the separation of the two slits.

We neglect this effect in the present discussion. The mathematics of the double-slit interferometer is discussed in an e-appendix available as a downloadable pdf file at our website, http://www.emmynoether.com.

8. Ibid.

9. Emily Dickinson, from "Part IV: Time and Eternity," in *The Complete Poems of Emily Dickinson*, with an introduction by her niece, Martha Dickinson Bianchi (Boston: Little, Brown, 1924).

10. Joseph Fraunhofer, see *The Encyclopedia of Science*, http://www.daviddarling.info/encyclopedia/F/Fraunhofer.html (accessed May 21, 2010).

11. Michael Faraday: Faraday and Maxwell are the two pillars of classical electrodynamics; see Alan Hirschfeld, *The Electric Life of Michael Faraday*, 1st ed. (New York: Walker, 2006).

12. Basil Mahon, *The Man Who Changed Everything: The Life of James Clerk Maxwell* (Hoboken, NJ: Wiley, 2004).

CHAPTER 4

1. The radiated energy by a hot object can be transported in three ways: (1) *conduction*, by direct contact between two or more objects, as in the case of an egg in direct contact with hot water; (2) convection, as by contact of a hot object with the air, heating the surrounding air, which then carries the energy away by motion of the air, as in the principle of forced air heating of your home on a cold night; (3) *radiative*, where the radiated energy is generally electromagnetic radiation, as in the case of the toaster wires with their red glow. The electromagnetic radiation is then barely visible and invisible light, but at extremely high temperatures, as in nuclear explosions, there will be radiated very high-energy electromagnetic radiation such as x-rays and gamma rays. The electromagnetic radiation is present within the object, bouncing around, emitted, and reabsorbed, helping to maintain the thermal equilibrium until it is finally radiated away from the surface of the object. Cooling of the human body is a mixture of these effects, accomplished by the evaporation of water (sweat) at the surface of the skin. Here, water is spontaneously converted on the surface of the skin from a liquid state to a water vapor in the air, provided the air is not too humid. The evaporation (transformation from liquid to gas) consumes energy and this comes by conduction from the skin, producing the cooling effect. The heat is then convected away.

2. For fireworks colors, see http://chemistry.about.com/od/fireworkspyrotechnics/a/fireworkcolors.htm; http://www.howstuffworks.com/fireworks.htm

(accessed April 15, 2010); see also *The Teacher's Domain*, http://www.teachers domain.org/resource/phy03.sci.phys.matter.fireworkcol/ (accessed January 1, 2010).

3. For *Rigel*, see http://en.wikipedia.org/wiki/Rigel (accessed January 1, 2010).

4. The basic unit of temperature used by physicists is the kelvin (K). *Absolute zero* is where matter contains no thermal energy and is defined as being precisely 0 K (corresponding to –273.15 °C). Even at absolute zero a system will have *quantum zero point energy*, since motion can never be zero in quantum physics. One unit of kelvin has precisely the same magnitude as a one-degree increment on the Celsius scale. Celsius zero is approximately the freezing point (triple point) of water, which corresponds to 273.15 kelvins (or 0 K = –273.15 °C). Wikipedia gives a more precise definition of thermal measurement systems, with links to references and conversion formulas: http://en.wikipedia.org/wiki/ Temperature (accessed January 1, 2010).

5. Josiah Willard Gibbs (1839–1903): Muriel Rukeyser, *Willard Gibbs: American Genius* (Woodbridge, CT: Ox Bow Press, 1942). Raymond John Seeger, *J. Willard Gibbs: American Mathematical Physicist Par Excellence* (Oxford, NY: Pergamon Press, 1974); L. P. Wheeler, *Josiah Willard Gibbs: The History of a Great Mind* (Woodbridge, CT: Ox Bow Press, 1998). The earliest hints of the quantum revolution can be found in the work of J. Willard Gibbs in the 1860s, as he was formulating the science of thermodynamics. An essential concept here is "entropy," which is the measure of all possible states of motion that a system of atoms, such as those comprising a gas, has for any specified volume, number of atoms, total energy, and so on. The key notion of "equilibrium" is a kind of stability of such a system—a room full of air can be in an equilibrium, an apparently static unchanging state, even though the atoms are swirling about and colliding with one another at short, unobserved distance scales. Gibbs realized that if the room is slowly partitioned into two equal halves, the static, unchanging equilibrium state of the gas should not change. For this to be so, the entropy of the half room should be one-half of the total entropy of the full room (entropy is "extensive"). If this were not so, Gibbs determined, there would develop fake pressures and temperatures (disequilibrium) associated with the act of merely partitioning the system. But the classical theory didn't give this result. The problem arises from the fact that, in classical theory, every atom is *distinguishable in principle* from every other atom. (A talented engraver could, in principle, engrave a different name on each helium atom: Rick, Katie, Graham, Mary, Ron, Don, etc.) To solve this problem (which is known as the "Gibbs paradox"), Gibbs had to introduce a "fudge factor" into his calculations that implies that all atoms of the same species are indistinguishable (e.g., all oxygen molecules are

equivalent; all nitrogen molecules are equivalent; oxygen molecules are distinguishable from nitrogen molecules; etc.). Even as a matter of principle, all atoms/molecules/particles of the same kind (e.g., helium atoms) are indistinguishable from one another.

This was a shocking result to people who understood the deep philosophical underpinnings of classical physics, but there were very few such people back then. The great James Clerk Maxwell, the formulator of classical electromagnetic theory and the leading scientist of the day, was an avid reader and user of the works of Gibbs, and he championed him as a scientist of equal par. Gibbs might otherwise have remained an obscure eccentric and unknown American physicist and might never have been accorded the fame he deserved. In any case, it wasn't clear at the time what Gibbs's fudge factor was telling us—perhaps it was a "mere issue" in the mathematical definition of the concept of entropy. It is with our 20/20 hindsight that we see this was the smoldering beginning of the quantum revolution. The enforced indistinguishability of identical particles is foundational to quantum theory and has profound consequences for the physical world (such as how electrons fill the orbitals of the atoms). Seeing this would take another sixty years of thinking and being confused. (See the appendix.)

6. Ludwig Boltzmann (1844–1906): David Lindley, *Boltzmann's Atom: The Great Debate That Launched a Revolution in Physics* (New York: Free Press, 2001); John Blackmore, ed., *Ludwig Boltzmann—His Later Life and Philosophy, 1900–1906, Book One: A Documentary History* (Dordrecht, Netherlands: Kluwer, 1995); Stephen G. Brush, *The Kind of Motion We Call Heat: A History of the Kinetic Theory of Gases* (Amsterdam: North-Holland, 1986). Boltzmann was a staunch advocate of the atomic theory and one of the visionaries who provided the quantum theory with what has become its major tool set. He invented the concept of "phase space," related to "entropy," which measures the number of available physical states a system of waves, with various wavelengths, can occupy. This instructs us how to calculate the resulting radiation pattern from, for example, a blackbody. This concept is foundational to the quantum theory and is present in every application of it to describe the world; it even plays a fundamental role in string theory today. Boltzmann suffered from depression, probably bipolar disorder, and died by his own hand at age sixty-two.

7. J. L. Heilbron, *Dilemmas of an Upright Man: Max Planck and the Fortunes of German Science* (Cambridge, MA: Harvard University Press, 2000); Max Planck, *Scientific Autobiography and Other Papers* (Philosophical Library, 1968).

8. Albert Einstein: There are far too many references to itemize here. For the quotation and an excellent biography, see Walter Isaacson, *Einstein: His Life and Universe* (New York: Simon & Schuster, 2007), p. 96. Perhaps the best biography is Abraham Pais, *Subtle Is the Lord: The Science and the Life of Albert Ein-*

stein (New York: Oxford University Press, 2005); see also A. Pais, "Einstein and the Quantum Theory," *Review of Modern Physics* 51, no. 4 (1979): 863–914.

9. W is called the "work function" of the metal. When the frequency f is greater than F, on the other hand, we have electron emission, and the electron, after paying the surface barrier toll, is left with an energy equal to the amount of energy it absorbed, decreased by the toll. In formula, the emitted electron's energy is $E = hf - W$. In English, the equation states that the electron's energy, after escaping from the metal surface, is equal to the energy the electron swallowed from the photon (hf) minus the toll needed to escape (W). Over the next years, this equation would be tested carefully by dozens of experimental scientists. It was correct! One can now look up the toll, what we call the "work function of the metal," W, tabulated in reference books. W, the surface energy toll, depends on the atomic structure of the metallic substance.

10. See *How Quantum Dots Work*, http://www.evidenttech.com/quantum dots-explained/how-quantum-dots-work.html (accessed May 21, 2010).

11. Each photon, which has energy, must also have a momentum of $p = E/c = hf/c$. This was confirmed by Compton's experiment that reveals individual photons (x-rays) colliding with individual electrons like relativistic billiard balls. See http://en.wikipedia.org/wiki/Compton_effect, and a bibliographical article at http://nobelprize.org/nobel_prizes/physics/laureates/1927/compton-bio .html (accessed May 21, 2010).

12. Ibid.

13. Ibid.

14. This discussion follows Richard P. Feynman's masterful Messenger Lectures, delivered at Cornell University in 1964; Richard P. Feynman, *The Character of Physical Law* (Cambridge, MA: MIT Press, 2001); see also R. P. Feynman, *Six Easy Pieces, Essentials of Physics by Its Most Brilliant Teacher* (Basic Books, 2005).

15. David Wilson, *Rutherford, Simple Genius* (Hodder & Stoughton, 1983); Richard Reeves, *A Force of Nature: The Frontier Genius of Ernest Rutherford* (New York: W. W. Norton, 2008). Good with his hands (unlike his mentor J. J. Thompson) and contemptuous of the head-in-the-clouds theoretical physicists, Rutherford was famous among his postdocs for acerbic quotes like the following: "Oh, that stuff [Einstein's relativity]. We never bother with that in our work."

16. Jan Faye, *Niels Bohr: His Heritage and Legacy* (Dordrecht, Netherlands: Kluwer Academic Publishers, 1991); see also the Niels Bohr Wikipedia article http://en.wikipedia.org/wiki/Niels_Bohr (accessed January 1 2010).

17. Oscar Wilde, "In the Forest," 1881, from *Charmides and Other Poems*, public domain (available online).

CHAPTER 5

1. Charles Enz and Karl von Meyenn, *Wolfgang Pauli: A Biographical Introduction, Writings on Physics and Philosophy* (Berlin: Springer-Verlag, 1994). C. P. Enz, *No Time to Be Brief: A Scientific Biography of Wolfgang Pauli*, rev. ed. (New York: Oxford University Press, 2002); see also David Lindorff, *Pauli and Jung: The Meeting of Two Great Minds* (Wheaton, IL: Quest Books, 2004).

2. Bohr realized, in 1911, that if the motion of an electron is like that of a wave, then the distance it travels, through a complete cycle of its orbit (the circumference of the orbit), must be a distinct number of *quantum wavelengths* of the electron's motion viewed as a wave. This, Bohr argued, is related, through Planck's constant, to the magnitude of the electron's momentum in its orbit. That is: the momentum of electron equals Planck's constant, *h*, divided by the quantum wavelength. The key to the atom is that this must match: the orbital circumference equals an integer number times the wavelength. Therefore, the electron's momentum can only take on certain special values that are related to the size of its orbit. This is how a musical instrument works—only certain distinct wavelengths of sound can be produced from a brass tube of a size, or a drum heard of a given diameter, or a string of a given length.

3. The binding energy of the electron in particular states of the atom might be 6.1 eV, 9.2 eV, 10.5 eV, and so on. An electron volt is a minuscule unit of energy that is useful at the atomic and subatomic scales. One electron volt (eV) is the energy that a single electron acquires if it falls through a voltage difference, in an electric circuit, of one volt. The "joule" is the unit of energy in the meter-kilogram-second system. It is often convenient to use a much smaller energy unit, such as the electron volt, for small things. The conversion shows how tiny this is, 1 joule = $6.24150974 \times 10^{18}$ eV. To give a sense of scales of energies, consider ordinary combustion. If we burn carbon, combining a carbon atom, C, and an oxygen molecule, O_2, we produce CO_2, and we get out about $E = 10$ eV in energy (in photons). In nuclear fission, a ^{235}U nucleus will typically convert to lighter nuclei (such as yielding about 200 MeV per fission). In nuclear fusion, we can combine a hydrogen nucleus (proton) with a deuterium nucleus (proton + neutron) to produce a helium isotope (two protons and one neutron) with a release of 14 MeV.

4. Frank-Hertz experiment, see http://hyperphysics.phy-astr.gsu.edu/hbase/frhz.html or http://spiff.rit.edu/classes/phys314/lectures/fh/fh.html (accessed May 21, 2010).

5. The conservation of momentum is a *vector equation*: $m_1, \vec{v}_1 + m_2\vec{v}_2 = m_1\vec{v}_1' + m_2\vec{v}_2'$ for, for example, a pair of billiard balls of masses m_1

and m_2 with initial velocities $(\vec{v}_1, \vec{v}_2)$ and final velocities $(\vec{v}_1', \vec{v}_2')$. Momentum conservation is a consequence of the fact that the laws of physics do not depend on where you are in space. This is an example of "Noether's theorem"; see Leon M. Lederman and Christopher T. Hill, *Symmetry and the Beautiful Universe* (Amherst, NY: Prometheus Books, 2004).

6. Louis de Broglie: Nobel Prize Speech (1929). See http://www.spaceandmotion.com/Physics-Louis-de-Broglie.htm (accessed May 21, 2010), and the Nobel Prize Biography, http://nobelprize.org/nobel_prizes/physics/laureates/1929/broglie-bio.html (accessed May 21, 2010). Although Bohr had essentially applied the notion that the electron was a wave to obtain his theory of the atom, he somehow considered this to be a feature of bound electrons in their orbits and didn't generalize the idea to untrapped, free electrons propagating through space. If an electron has a wave associated with it, what is its wavelength? Guided by Einstein's special relativity and Planck's relation of energy to wavelength, de Broglie suggested that a particle's wavelength would depend on both its mass and its velocity—in short, on its momentum. Momentum is defined as mass times velocity, or $p = mv$. De Broglie's inspired insight was that a particle must have a wavelength (call it λ) equal to Planck's constant h divided by its momentum p, or $\lambda = h/p$ (obviously, such a quantum idea has to introduce the quantum theory's logo, h). De Broglie realized that the faster the particle moves, or the more momentum the particle has, the smaller its associated wavelength must be.

7. Ibid.

8. Ibid.

9. David C. Cassidy and W. H. Freeman, *Uncertainty: The Life and Science of Werner Heisenberg* (1993); Arthur I. Miller, ed., *Sixty-Two Years of Uncertainty: Historical, Philosophical, and Physical Inquiries into the Foundations of Quantum Mechanics* (New York: Plenum Press, 1990).

10. Ibid.

11. See also *Werner Heisenberg Austolang: Helgoland* at http://www.archiv.uni-leipzig.de/heisenberg/Geburt_der_modernen_Atomphysik (accessed January 1, 2010).

12. Precisely, $xp - px = ih/2\pi$ is the defining "commutator" of quantum mechanics in Heisenberg's language. The correspondence with Newton follows fairly naturally: in the world of large things, like an elephant or a locomotive or specks of sand, we can get away with using ordinary commuting numbers to measure the position x (in, let's say, meters), and if an elephant is charging, it has a momentum $p = Mv$ (M is the elephant's mass in kilograms; v is his velocity in meters per second). We can use ordinary numbers to define the classical ele-

phant and $xp - px = 0$ because the description of that macroscopic elephant is insensitive to Planck's h. But for tiny particles—electrons, atoms, photons—this is no longer true. In fact, it is even true for the elephant that $xp - px =$ (a number) times h, but there is no experiment sensitive enough to see this for an elephant . . . it requires the delicacy of individual atoms to see that nature doesn't use everyday mathematics to describe electrons! And, if this isn't enough for you, the "number" that multiplies h in the above expressions is $i/2\pi$, where i is the square root of minus one! We have definitely landed in Oz.

The following example is taken from Lederman and Hill, *Symmetry and the Beautiful Universe*," figure A3, p. 303. Take a copy of a book—any book will do. We can perform simple rotations on the book. We can think of an imaginary coordinate system with the book placed at the origin. Now rotate the book through 90° around the imaginary x-axis. We always rotate using the "right-hand rule." Call this operation "**a.**" Now follow this rotation by another 90° rotation around the imaginary y-axis. Call this operation "**b**" Look at where the book ends up. This is the result of **a** × **b**. Now go back to the initial position of the book; next, rotate first along the y-axis (**b**) and follow it by a rotation along the x-axis (**a**) and note the position of the book (**b** × **a**). Is **a** × **b** equal to **b** × **a**? The answer is an emphatic no! The order in which we perform rotations matters! The noncommutativity is a property of the rotations themselves, a property of nature and not of the object we are rotating.

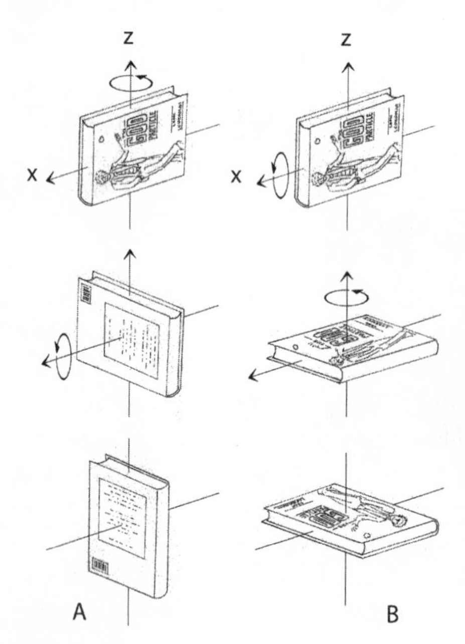

FIGURE 38: Start with a book in the initial position. Rotate by 90° about the z-axis, then 90° about the x-axis. We get the book to position (A). Repeat the experiment: return to the initial position, rotate first 90° about the x-axis, and then 90° about the z-axis. We get the book into a different position, (B). The rotations do not commute, that is: $X \times Z \neq Z \times X$. (Illustration by Shea Ferrell.)

Heisenberg reasoned that a measurement of the position of an object, **x**, followed by the measurement of its momentum, ***p***, would yield a different result than doing things in the opposite order, that is, measuring first the momentum, ***p***, followed by the position, ***x***. In creating a quantum description of this, he found that the act of measurement was like ordinary multiplication of these symbols, and again, ***a*** times ***b*** can generate a different product than ***b*** times ***a***. To be precise, if ***x*** is the position of an electron, and ***p*** is the momentum, Heisenberg discovered that ***xp*** does not equal ***px***. Now of course, this doesn't happen in Newtonian physics. Measuring position followed by momentum is ***xp*** and measuring momentum followed by position is ***px***, and the two are always equal. But quantum physics has at its heart, so reasoned Heisenberg, the central feature that measuring position necessarily messes up the momentum of a particle, and vice versa, measuring the particle's momentum necessarily messes up the position of the particle. The difference ***xp*** – ***px*** is very, very small . . . associated with quantum effects and so must be about the size of Planck's constant, h.

13. If the uncertainty in momentum is large, then the momentum itself would grow as large as the uncertainty, $\Delta p \geq \hbar/2\Delta x$. The kinetic energy (energy of motion) is determined by momentum, $KE = p^2/2m$, and it, too, would therefore get large as $(\Delta p)^2/2m \geq \hbar^2/2m(\Delta x)^2$, where m is the electron's mass. This would become much larger than the negative potential energy, $PE = -e^2/x \cong -e^2/\Delta x$. So the total energy $KE + PE$ actually increases as we squeeze the electron closer to the nucleus—the atom has a ground state and is stable. This effect is called "Schrödinger pressure," and it generally shows how nonrelativistic quantum physics resists the collapse of a system. Schrödinger pressure can be overcome, however, in the relativistic limit. There the energy is directly related to the momentum, at high momentum, $E \cong pc \cong \hbar c/2\Delta x$. Now the potential energy for a system with inverse square law force, $PE \cong -k/\Delta x$, can win. Hence, enormous stars can collapse when their interiors become relativistic, and black holes can form.

14. Walter J. Moore, *Schrödinger: Life and Thought* (Cambridge, MA: Cambridge University Press, 1992); see also J. J. O'Connor and E. F. Robertson (on Schrödinger) online at http://www-gap.dcs.stand.ac.uk/~history/Mathematicians/Schrodinger.html; K. von Meyenn, "Pauli, Schrödinger and the Conflict about the Interpretation of Quantum Mechanics," in *Symposium on the Foundations of Modern Physics* (Singapore, 1985), pp. 289–302. The Oppenheimer quote is also referenced in Dick Teresi, "The Lone Ranger of Quantum Mechanics," in the *New York Times* book review of *Schrödinger: Life and Thought*, by Walter J. Moore (above), January 9, 1990. For example, if x is a position along the direction of motion of the wave, and t is the time, then we can describe a particular traveling

wave by a sinusoidal function of the form: $\Psi(\vec{x},t) = A\cos(kx - \omega t)$. When plotted at any time, t, this is a *wave train*, and as the time t increases, the wave train moves to the right. The quantity k is called the *wave number* and the quantity ω is called the *angular frequency* of the wave. These things are often related to the usual "cycle-per-second" *frequency* $f = \omega/2\pi$, and the *wavelength* $\lambda = 2\pi/k$. λ is the distance between two neighboring troughs or crests of the wave. f is the number of times per second that the wave undulates up and down through complete cycles at any fixed position x. In other words, if you think of the wave as a long freight train, then λ is the length of a boxcar and f is the number of boxcars per second passing in front of you as you patiently wait for the train to pass. A is called the *amplitude* of the wave and determines the height of the crests, for example, the distance from a trough to a crest is $2A$. The velocity of the traveling wave is $c = \lambda f = \omega/k$. This is usually written as a vector, and kx is usually written as $\vec{k} \cdot \vec{x}$ in three dimensions of space, to represent a wave traveling in the direction $\vec{k}$.

15. "A Digression on Complex Numbers": The real numbers were probably discovered in early Mesopotamia in the West and ancient China in the East. It may seem peculiar that numbers must be "discovered," but that is, in fact, the case. We start with simple counting numbers, or the integers, 0, 1, 2, 3, . . . , which were discovered by counting sheep and money, but soon we discover the negative integers, –1, –2, –3, ... This happened when somebody invented "subtraction" and tried to subtract 4 from 3. The Greeks discovered the *rational numbers*, that is, those numbers that can be written as the ratio of two integers, such as 3/4 or 9/28. The Greeks also discovered the *prime numbers*, that is, any integer that cannot be evenly divided by an integer other than itself, such as 2, 3, 5, 7, 11, 13, 17, ... Hence $15 = 3 \times 5$ is not prime but *contains the prime factors* 3 and 5. In a sense, the primes are the "atoms" out of which all integers are built up by multiplication. The primes are of profound importance in mathematics and remain the focus of many ongoing studies of their properties even today. Numbers like $\sqrt{2}$ and π are *irrational numbers* and cannot be expressed as the ratio of two integers. Taken together, the positive and negative integers, rationals and irrationals, make up the collection of *real numbers*.

The Arabs invented algebra and began to solve problems like $x^2 = 9$, finding two solutions, $x = 3$ and $x = -3$. Soon thereafter they discovered the *imaginary numbers*. For example, we may want the solution to the problem $x^2 = -9$. There is no real number that solves this equation. We therefore invent a new number, called i, which is *defined* as $i = \sqrt{-1}$, or $i^2 = -1$. There are thus two solutions to our equation, $x = 3i$ and $x = -3i$ We can then build new numbers of the form $z = a + bi$, where a and b are both real. These are called *complex numbers.* We define the *complex conjugate* of x to be $z^* = a - bi$, and the *magnitude* of z to be $|z| = \sqrt{zz^*} | = | \sqrt{a^2 + b^2}|$. The imaginary numbers represent a *second dimension*, that is, a perpendicular axis, to the conventional real number line. This leads to the *complex plane*, in which the x-axis is the ordinary real number line and the y-axis is the set of all real numbers multiplied by i. Complex numbers are *vectors* in the complex plane.

A theorem of fundamental importance connects *exponentials of imaginary numbers* to complex numbers through *trigonometric functions*: $e^{i\theta} = \cos(\theta) + i\sin(\theta)$. The proof of this is often relegated to a course in calculus, but it can, in fact, be proven using just the general properties of exponentials and the "addition theorems" for trigonometric functions. Using this result, any complex number can be written as $z = \rho e^{i\theta}$, where ρ and θ are real. Then $|z| = |\sqrt{zz^*}| = |\rho$. This is a polar coordinate representation of the complex plane.

Is there really a physical significance to the use of the complex numbers in physics equations? In quantum physics, it makes no sense not to accept the fact that there *really are complex numbers* and the *wave function really is a complex valued function of space-time*. The fact is, in the mathematics of quantum mechanics the square root of negative one, i, *plays a fundamental role*. Quantum mechanics is essentially a theory of the square root of probability, and i will naturally arise in such a construct. Nature evidently reads books on complex numbers

16. At this point many students say: "Surely you jest! Don't you mean that you are merely using complex numbers as a kind of mathematical tool or convenience, like people do in electrical engineering, and there really is no physical significance to the use of the complex numbers in physics equations?" To which we answer: "No! We do not jest!" In quantum mechanics there *really are complex numbers*, and *the wave function really is a complex valued function of space-time*. Now, of course, we could reduce everything to pairs of real numbers and painfully do all the math without ever talking about the combination of the square root of minus one, i, but there is no advantage to doing so. That would be like talking in a circumspect way about a horrible social disease at a cocktail party without ever using the actual name of the disease, but everyone would still understand what we were really talking about, and someone might sooner or later blurt it out. The fact is, in the mathematics of quantum mechanics the square root of negative one, i, *plays a fundamental role*. We don't know why, but we know it is true. So what does a quantum particle wave function look like?

Using Schrödinger's wave equation, we would find that a traveling particle is a traveling wave with a wave function that looks something like this

$$\Psi(\vec{x},t) = A\cos(\vec{k}\cdot\vec{x} - \omega t) \text{ where } |\vec{k}| = 2\pi / \lambda \text{ and } \omega = 2\pi$$

17. A string on a violin or guitar vibrates like the wave function of a bound electron. The string is pinned down at two places, one by the *bridge* and the other by the *nut* at the end of the neck of the instrument. When we pluck a string, it vibrates, producing a musical note. The vibrations of the string are *trapped* or *standing waves*. Indeed, if the length of the string were infinite, we could pluck the string and send a traveling wave down the string, off to infinity, representing a free traveling particle in empty space in quantum mechanics. But our string has a fixed length, spanning the nut to the bridge, denoted by L, typically about one meter for the average guitar and a foot for a violin.

If we lightly pluck the string at its midpoint, we excite the *lowest mode of vibration* of the string. This corresponds to the lowest quantum energy state of motion of the electron trapped in the ditch. This mode of vibration has a wavelength that is $\lambda = 2L$ (pronounced "LAM-dah"), which means the length L is just *half of an entire wavelength* (i.e., there is only one crest or one tough at the peak of the oscillation, while a full wavelength would have both a crest and a trough). This is the *lowest mode* or *lowest energy level* or the *ground state* of the system, corresponding to the lowest note of the plucked string.

Now, we consider the *second mode* of oscillation of the string. This mode has a wavelength that is now $\lambda = L$. That is, we can have both a crest and a trough within the full distance $\lambda = L$. You can actually excite the second mode on a real guitar string, with a little patience, by lightly holding your finger to the midpoint of the string while plucking it halfway between your finger and the bridge, then quickly removing your finger. The finger ensures that the center of the string doesn't vibrate, which we see is a feature of the second mode of oscillation (such stationary points are called *nodes* of the wave function). This produces a pleasant, and somewhat harplike, angelic tone, one octave above the lowest mode. Because this has a shorter wavelength, the second mode of the quantum particle has a higher momentum and therefore a higher energy than the lowest mode.

If we shine a photon on our electron with just the right amount of energy, we can accelerate the electron and cause it to hop into the second mode, or *the first excited quantum state* of the system. Likewise, the electron can radiate a photon and jump down from this state into the ground state. The next sequentially higher energy level is the third mode of vibration of the string, which has one and a half full waves, which means $\lambda = 2L/3$. This can be excited on the guitar by holding one's finger at one-third the length of the string below the nut and plucking at the midpoint of the bridge and finger, then quickly removing the finger—one should then hear a very faint angelic fifth note (if the string is tuned to C, this note is G in the second octave above C). This corresponds, therefore, to a still shorter wavelength and a correspondingly large momentum, and thus the energy is still larger.

Again, a photon hitting our electron with just the right amount of energy can accelerate the electron into this state from the other excited states. Or the electron can radiate a photon and jump down into the lower energy states from here. With the application of more energy, we can get the electron to hop to the fourth, fifth, sixth, and so on, energy levels, which each correspond to the higher modes of vibration of the guitar string. Eventually, the electron can get sufficient energy to escape the potential, and it then becomes a free particle (its wave function travels away from the scene). We say that the system has become *ionized.*

18. See note 14.

19. Nancy Thorndike Greenspan, *The End of a Certain World: The Life and Science of Max Born*, export ed. (New York: Basic Books, 2005); G. S. Im, "Experimental Constraints on Formal Quantum Mechanics: The Emergence of Born's Quantum Theory of Collision Processes in Göttingen, 1924–1927," *Archive for History of Exact Sciences* 50, no. 1 (1996): 73–101; For more on Born, see http://www-gap.dcs.st-and.ac.uk/~history/Mathematicians/Born.html (accessed January 1, 2010).

20. We will, in the text, ignore the fact that $\Psi(x, t)$ is complex and we'll write $\Psi(x, t)^2$ for the probability for simplicity. In fact, we mean the absolute square, $|\Psi(x, t)|^2$, which is a positive quantity reflecting the squared magnitude of the complex wave function. $|\Psi(x, t)|^2$ is the probability density, that is, $|\Psi(x, t)|^2\,dx$ is the local probability of finding the particle in a differential interval dx, assuming we are talking about one dimension of space. The probability of finding a real particle anywhere in a three-dimensional spatial volume, V, involves an integral (in calculus) over the volume at fixed time t, for example, in three space dimensions $\int_V |\Psi(\vec{x},t)|^2 d^3x$. For all the space occupied by the particle, this is 1. The Schrödinger equation makes the probability remain 1 for all space for all time. This is called "conservation of probability," or "unitarity," and it can be proven that it requires a special constraint on the theory (called "hermiticity of the Hamiltonian"). Hermiticity enforces the energies of physical states to be real numbers.

21. See note 19.

22. See note 15.

23. See note 20.

24. Heisenberg's equation is actually three-dimensional, representing similar statements about the location and momentum of a particle along any one of three space directions. Not surprisingly, Heisenberg's rules prevent us from resolving the wave–particle dilemma as revealed in the double-slit experiment.

CHAPTER 6

1. Heinz R. Pagels, *Perfect Symmetry: The Search for the Beginning of Time* (New York: Simon & Schuster, 1985).

2. *Apollo* 15 astronaut Cmdr. David Scott is seen dropping a feather and a hammer on the moon in this video: http://video.google.com/videoplay?docid=6926891572259784994# (accessed January 1, 2010).

3. See our e-appendix online at http:// www.emmynoether.com.

4. Erwin Schrödinger, *What Is Life? Mind and Matter* (Cambridge: Cambridge University Press, 1968).

5. James D. Watson, *The Double Helix: A Personal Account of the Discovery of the Structure of DNA* (New York: Touchstone, 2001).

6. Roger Penrose, *Shadows of the Mind: A Search for the Missing Science of Consciousness* (New York: Oxford University Press, 1996).

7. Michael D. Gordin, *A Well-Ordered Thing: Dimitry Mendeleev and the Shadow of the Periodic Table*, 1st ed. (New York: Basic Books, 2004); Dmitri Ivanovich Mendeleev, *Mendeleev on the Periodic Law: Selected Writings, 1869–1905*, edited by William B. Jensen (Mineola, NY: Dover, 2005).

8. The "atomic weight" or, equivalently, "atomic mass," of carbon is defined to be A = 12.00. Hydrogen has about one unit of mass, or A = 1. So why do we see 1.0079 on the modern charts compared to 1/12 of the mass of the carbon-12 atom? There are two reasons for this: (1) some of C's mass is binding energy of protons and neutrons in the nucleus, and (2) these masses are quoted for particular blends of *isotopes*, which are mixtures of the same atom (Z = proton number), but having different numbers of neutrons in the nucleus. Naturally occurring H in seawater is actually a mixture of pure D (deuterium). Isotopes are explained in this Wikipedia article at http://en.wikipedia.org/wiki/Atomic_weight (accessed January 1, 2010). The sequence in the lineup of atomic masses goes as follows: hydrogen (Z = 1, A = 1), helium (Z = 2, A = 4), lithium (Z = 3, A = 7), beryllium (Z = 4, A = 9), boron (Z = 5, A = 11), carbon (Z = 6, A = 12), nitrogen (Z = 7, A = 14), oxygen (Z = 8, A = 16), fluorine (Z = 9, A = 19), neon (Z = 10, A = 20), sodium (Z = 11, A = 23), magnesium (Z = 12, A = 24), aluminum (Z = 13, A = 27). As we can't do full justice to the subject in this limited space, we advise

that you consult the literature, much of which is on the web at http://en.wikipedia.org/wiki/Periodic_table and http://www.corrosion-doctors.org/Periodic/Periodic-Mendeleev.html (both accessed January 1, 2010).

9. Ibid.

10. See note 7.

11. Lithium and sodium fires in reaction with water: http://www.youtube.com/watch?v=oxhW7TtXIAM&feature=related; http://www.youtube.com/watch?v=Jw9p-5t8 wWY&feature=related (accessed April 30, 2010). So what's going on here? First, the water molecule is broken down at the surface of the lithium into parts $H_2O \rightarrow H + OH$ (water likes to do this; the OH is called "hydroxyl" and usually attaches to an extra electron to have a net negative charge, while the H is usually just a bare proton that lost its electron to a hydroxyl). (An "ion" is an atom or small compound that has attached to an extra electron or lost an electron to become electrically charged; the protons, or H^+s, in liquid water actually attach to H_2Os to make "hydronium ions," H_3O^+; too many extra hydronium ions floating around in water [or too few OH ions], and you have an "acid," too many extra OHs [too few hydroniums] and you have a "base.") Lithium rapidly steps into a place where one of the hydrogens was, forming LiOH (lithium hydroxide). This discards the extra hydrogen, whose place Li took. This extra H furiously bubbles off the surface of the water near the block of Li, but so much heat is released by this interaction that there is often ignition. To appreciate the violence of this reaction, please see the above videos. (We actually have personal experience with this. Believe us, it can be ferocious.)

12. Historically, Mendeleyev didn't know about He, Ar, and so on, the "noble gases," so his periodic table had second and third rows containing only seven elements. See http://www.elementsdatabase.com/ and http://www.bpc.edu/mathscience/chemistry/images/periodic_table_of_elements.jpg (accessed May 21, 2010).

13. We mean here the *magnitude of the momentum*, because the trapped wave is not in a state with a single definite momentum, that is, it is not a traveling wave. The standing wave has two values of the momentum at any instant, one positive and one negative, but with a definite common magnitude. The wave function for the lowest mode is just the shape of the vibrating guitar string in space, oscillating in time. The shape of the lowest mode is $(\pi x / L$. The exact form of the wave function involves, as it must, complex numbers and can be written as $\Psi(x, t) = A\sin(\pi x / L)e^{i\omega t}$, where $\omega = 2\pi E/ h$ is the angular frequency. The probability of finding the electron somewhere between $x = 0$ and $x = L$ is therefore $|\Psi(x, t)|^2 = A^2 \sin^2(\pi x / L)$. In fact, if the probability is unity to find the electron somewhere in the interval $0 \leq x \leq L$, then we find that $A = 1\sqrt{2L}$.

14. These are none other than the squares of Schrödinger's wave functions, $|\psi|^2$. Where the fuzzy cloud is darkest, the electron is most likely to be; where the cloud is faint and fades away, the electron is unlikely to be. If you tamper with the atom by performing a measurement of the electron's position, let us say, by shooting a high-energy photon (a gamma ray) into the cloud to strike the electron if it is at that point in space, your "hits," repeated many times, would cluster where the "cloud" was most dense and would be very rare where the cloud was tenuous and faint. See http://en.wikipedia.org/wiki/Atomic_orbital (accessed January 2, 2010).

15. George Gamow, *Thirty Years That Shook Physics* (New York: Dover, 1985).

16. Incidentally, ordinary molecular hydrogen is a mixture of states where the spins of the protons are counteraligned to form a singlet (up, down)-(down, up) [parahydrogen], or it can be in any of three "triplet states" (up, up); (up, down) + (down, up); (down, down); [orthohydrogen]. See http://en.wikipedia.org/wiki/Orthohydrogen (accessed January 1, 2010).

CHAPTER 7

1. John Rigden, *I. I. Rabi: Scientist and Citizen* (Cambridge, MA: Harvard University Press, 2001).

2. Electron spin was first observed in the Stern-Gerlach experiment. See http://library.thinkquest.org/19662/low/eng/exp-stern-gerlach.html and http://plato.stanford.edu/entries/physics-experiment/app5.html (accessed May 21, 2010).

3. Ibid.

4. Pascual Jordan, *Physics of the Twentieth Century* (Davidson Press, 2007).

5. See chapter 1, note 3.

6. William Shakespeare, *Hamlet*, act 1, scene 5, lines 166–67.

7. M. Beller, "The Conceptual and the Anecdotal History of Quantum Mechanics," *Foundations of Physics* 26, no. 4 (1996): 545–57; L. M. Brown, "Quantum Mechanics," in *Companion Encyclopedia of the History and Philosophy of the Mathematical Sciences*, edited by I. Grattan-Guinness (London, 1994), pp. 1252–60; A. Fine, "Einstein's Interpretations of the Quantum Theory," in *Einstein in Context* (Cambridge: Cambridge University Press, 1993), pp. 257–73. M. Jammer, *The Philosophy of Quantum Mechanics: The Interpretations of Quantum Mechanics in Historical Perspective* (New York, 1974). Jagdish Mehra and Helmut Rechenberg, *The Historical Development of Quantum Theory* (New York; Berlin, 1982–1987).

8. "Mixed States in Quantum Theory": Put another way, consider a tree in the forest. In classical physics we can say that the tree is standing, and it eventually topples over, following a definite path specified by the angle θ, the angle of the tree from vertical, as a function of time. Initially, the standing tree has $\theta = 0$ degrees. Over time, the tree starts to tilt and gradually θ becomes 10 degrees, then 20 degrees, and finally, the tree falls down, with a crash, when θ is then 90 degrees from the vertical for the toppled tree. We can go into the forest and measure (or "observe") θ at any given time. We could even post a video camera to measure θ and record the angle as it changes gradually in time, from a standing healthy tree with $\theta = 0$, to toppled with $\theta = 90°$. We could make a graph showing how $\theta(t)$ moves from $\theta(0) = 0$ initially when $t = 0$, to $\theta(T) = 90°$ at some later time, $t = T$. This is classical physics and our commonsense "classical" intuition.

Now contrast this with the quantum theory. We consider the tree as a quantum mechanical system, such as an atom, to make the analogy. We build a "tree detector." Our detector can only observe the tree to be in one of two quantum states: either tree is standing vertically (spin-up), which we denote by the quantum state: (tree up); or the tree has fallen (spin-down): (tree down).

But what about the actual act of the falling of the tree? Here the quantum theory allows us to construct a new state *in which the tree has neither fallen nor remained standing vertically*, called a "mixed state":

$$a \text{ (tree up)} + b \text{ (tree down)}$$

In quantum physics a and b are *complex numbers* (see note 5). Quantum physics enforces a rule (called "unitarity") that at any time $|a|^2 + |b|^2 = 1$. According to quantum theory, $|a|^2$ is the probability that the tree is standing, and $|b|^2$ is the probability that the tree has fallen down. This simply states the obvious fact that the total probability that the tree is either still standing or has fallen is 1. The probabilistic interpretation of quantum theory would make no sense if unitarity were not true (unitarity is called the "conservation of probability"). Initially the tree is in the "pure state" of standing vertically in the forest, and this means that $a = 1$ and $b = 0$. But as time goes forward, the laws of quantum theory allow a and b to change, always preserving the rule $|a|^2 + |b|^2 = 1$.

Now, suppose we go back later and observe the tree with our "tree detector." When we observe the tree, we may find that it is still standing (with the probability $|a|^2$). But, according to Bohr, this mere act of observation resets the quantum state back to (tree up); that is, now $a = 1$ and $b = 0$ again . . . simply because we measured the system and discovered it was "tree vertical." The act of measurement changes the physical system from the mixed state back to the pure state. This is one of the bizarre aspects of the quantum theory—the act of measurement disturbs and changes the quantum state in a fundamental way. Notice how different this is from classical physics in which we can observe the tree listing, for example, at thirteen degrees from the vertical, then depart, leaving the tree still with $\theta = 13°$, without influencing the tree at all.

But now we return to the quantum tree with our detector and we observe that the tree has fallen (with a probability of $|b|^2$). This act of observation now resets the quantum state to (tree down), where now $a = 0$ and $b = 1$.

The quantum state before the observation can be the mixture of both. The observation of tree fallen or tree standing is determined only with a certain probability and then defines a new state of the tree. In fact, anything (e.g., an alien passerby) that could in principle have made the observation, even if it isn't us, defines the new state of the tree. If there's no alien, and no us, nothing to disturb the tree and carry information away as to what the state is, then the tree remains in its mixed state; but an alien observation, or our observation, or an electronic automated observer resets the state into one of the definite possibilities, up or down.

In Heisenberg's language for the Schrödinger cat we have a mixed state: a (cat is alive) + b (cat is dead). In this case, the states have probabilities of a^2 for cat alive and b^2 for cat dead, and, of course, $|a|^2 + |b|^2 = 1$. Initially the cat is alive, so $a = 1$ and $b = 0$, but as time goes on, a gets smaller and b gets larger. But is the cat alive or dead? According to quantum theory, we don't know until we take a peek. The alive/dead mixture changes the minute you look in the box. If the cat is still alive, the quantum state resets to (cat alive) with $a = 1$ and $b = 0$. But, if to our horror, the cat has expired, the state sets to (cat dead) with $a = 0$ and $b = 1$.

9. Nathan Rosen, et al., eds., *The Dilemma of Einstein, Podolsky and Rosen—60 Years Later: An International Symposium in Honour of Nathan Rosen*, Haifa, March 1995, *Annals of the Israel Physical Society* (Institute of Physics Publishing, 1996); see also http:/en.wikipedia.org/wiki/EPR_paradox (accessed April 30, 2010). M. Paty, "The Nature of Einstein's Objections to the Copenhagen Interpretation of Quantum Mechanics," *Foundations of Physics* 25, no. 1 (1995): 183–204.

10. Our explanation of Bell's theorem here is adapted from the delightful website http://www.upscale.utoronto.ca/PVB/Harrison/BellsTheorem/Bells Theorem.html (accessed January 1, 2010) and provides a simple, special case example of Bell's theorem, which we have embellished here.

11. Ibid.

12. John Bell: http://www.americanscientist.org/bookshelf/pub/john-bell -across-space-and-time; see also http:/en.wikipedia.org/wiki/John_Stewart _Bell (accessed January 1, 2010).

CHAPTER 8

1. Electrons in a good electrical conductor move slowly compared to the speed of light, c. The electrons in the hydrogen atom, for example, in their quantum orbitals, have velocities of order 0.3 percent of the speed of light. Deep in the interior orbitals of larger atoms, the electron speeds begin to approach 10 percent of c, and the inner transitions emit x-rays and gamma rays. So then we have to start to worry about the effects of relativity. But these electrons are trapped deep within filled orbital shells. These inner configurations resemble the inert atoms, and the inner shell electrons do not interact chemically.

2. Einstein based special relativity on two principles: (1) *the principle of relativity*: all states of uniform motion (called "inertial reference frames") are equivalent for the description of physical phenomena; and (2) *the principle of the constancy of the speed of light*: all observers will obtain the same value for the speed of light in any inertial reference frame.

3. Relativity modifies the relationship among energy, momentum, and mass of a particle:

Einstein's relationship between energy and momentum: $E^2 - p^2c^2 = m^2c^4$

The energy of a particle is thus $E^2 = m^2c^4 + p^2c^2$, and we then have to take the square root of this mathematical expression for E. Taking the positive root for zero momentum, we get $E = mc^2$. For small momenta, we have approximately

$$E \approx mc^2 + \frac{p^2}{2m}$$

where the first correction is the Newtonian expression for the kinetic energy.

4. The conversion of one type of particle to another is not arbitrary and is subject to "selection rules" associated with the force or interaction that is causing the decay. For example, the proton cannot simply decay into an electron and a photon, because the proton has a positive electric charge and the electron has a negative charge. Perhaps a proton could decay into an antielectron (a positron of positive charge) and a photon, but this would require a new force in nature that causes the violation of "baryon number" and the violation of "lepton number" (the proton has baryon number 1, and no lepton number; the positron has lepton number –1 and no baryon number). We do believe that such interactions occur but are extremely feeble (in fact, such interactions occur in the Standard Model via very rare topological processes called "electroweak instantons").

The rate for the decay of the proton is so slow that the lifetime of a proton exceeds 10^{36} years!

5. See notes 2 and 3.

6. The negative energy particle would have energy given by

$$E = -\sqrt{m^2c^2 + p^2c^2}$$

This becomes more negative as the momentum, p, increases.

7. Please don't confuse negative energy particles with "tachyons." These are *hypothetical* particles that supposedly travel faster than the speed of light. Tachyons have an imaginary mass and thus satisfy a relationship like $E^2 = -m^2c^4 + p^2c^2$, which is different by the minus sign in the mass-squared term. There is no viable "theory" of tachyons as particles. Tachyons in quantum field theory usually imply that the vacuum is unstable, like a rock perched on top of a hill, where the tachyonic modes are "runaway modes," that is, representing the rock rolling down the hill, by which the entire vacuum becomes destabilized. Eventually at the "bottom of the hill" (the minimum of potential energy), the $-m^2c^4$ mass term becomes a normal positive expression, $+(m')^2c^4$ mass term, and the modes become normal particles.

8. Paul A. M. Dirac, *The Principles of Quantum Mechanics* (New York: Oxford University Press, 1982).

9. See note 6.

10. The positron is discussed with a picture of the positron track in Anderson's detector in this Wikipedia article at http://en.wikipedia.org/wiki/Positron; see also http://www.orau.org/ptp/collection/miscellaneous/cloud chamber.htm and http://www.lbl.gov/abc/Antimatter.html. See Dr. Christopher T. Hill lecture on antimatter in Fermilab's Saturday Morning Physics at http://www.youtube.com/watch?v=Yh1ZY1A2c5E&feature=watch_response (all accessed May 25, 2010).

11. Recently, the *DZero* experiment at the Fermilab Tevatron Collider may have produced the first evidence of new physics that could begin to unlock the secret of "why we exist." The key idea is that certain types of interactions can affect particles slightly differently than their antiparticles. These are called "CP-violating" interactions, and they are already known to exist, but up until now they have been observed to be too feeble to account for the amount of matter we see in our universe today. The new results pertain to a particle called a "heavy B_s meson" (which is an electrically neutral, spin–0 particle, with a mass of about 5 GeV/c^2, that is known to be composed of a b-quark and an anti-strange quark). This particle quickly "oscillates," changing its identity into its own antiparticle, the $\overline{B}_s$ meson. The $\overline{B}_s$ then quickly oscillates back into the B_s. These oscillations

happen rapidly and can occur many times before the heavy meson system radioactively decays in about a trillionth of a second. When the meson decays, it is either in the B_s phase or in the $\overline{B}_s$ phase of its oscillation. When it decays ("semileptonically") in the B_s phase, it produces a negatively charged "muon"; alternatively, when it decays while in the $\overline{B}_s$ phase, it produces a positively

charged "anti-muon." The B_s meson is typically produced together with a $\overline{B}_s$, and both of these particles go off on their merry way, oscillating back and forth into one another. So, when both of these have decayed, we would normally expect that positive and negative muons would be in a statistical balance. However, the CP-violating interactions can cause one of the two phases of the oscillation, say, the B_s phase, to persist slightly longer than the other, $\overline{B}_s$, phase. This means that there would be a slightly enhanced chance of detecting events in which a pair of negatively charged muons is produced (from both mesons being in B_s phase) over the anti-muons (both in the $\overline{B}_s$ phase). This slight excess of muons over anti-muons is what the *DZero* experiment at the Fermilab thinks it may be seeing. If so, the effect is about fifty times bigger than what theory, the Standard Model, predicts. As such, this could be the first hint of the breakdown of the Standard Model, as well as a hint of a new CP-violating force. This new force may have the necessary strength to explain why there is matter in our universe and no antimatter, hence, why we exist.

As of this writing, these results are very preliminary and they will require much patient work to confirm (such is science). Scientists around the world are very excited by the news, and more results should be forthcoming. Original scientific article: V. M. Abazov et. al., the DZero Collaboration, "Evidence for Anomalous Like-Sign Di-Muon Anomaly arXiv: 1005.2757 [hep-ex]"; see also a popular account: Dennis Overbye, "A New Clue to Explain Existence," at http://www.nytimes.com/2010/05/18/science/space/18cosmos.html (accessed May 17, 2010).

12. See Alexander Norman Jeffares, "William Butler Yeats," in *A New Commentary on the Poems of W. B. Yeats*, p. 51: "The Fish" first appeared in *Cornish* magazine, December 1898, with the title "Bressel the Fisherman."

13. If we do the sum of N negative integers (not counting zero), the result is the famous formula $-N(N + 1)/2$. This is called Gauss's formula. Allegedly, mathematician Carl Gauss derived it while a child in grade school, when his teacher told the class to add up all the numbers from 1 to 100. This is exactly how the sum would work for the Dirac sea if we lived in a world with one dimension of space and one of time.

14. Some older textbooks talk about the solutions to the relativistic wave equation for mesons and photons as having negative energies, but in fact the

quantum states corresponding to these solutions all have positive energy as dictated by the "Hamiltonian" of the full bosonic quantum field theory.

15. See "supersymmetry" at http://en.wikipedia.org/wiki/Supersymmetry (accessed March 10, 2010).

16. Any attempt to try to make the photon be the superpartner of the electron will fail, like Dirac's attempt to make the proton the antielectron, since the superpartner must also carry the same electric charge as the electron and the photon does not have *n* electric charge. There are some attempts to hide SUSY in a mysterious way so the vacuum energy problem is solved, but these mysteriously hidden SUSY theories are not very well defined and are not compelling. Yet, hope springs eternal for a clever new solution.

17. http://en.wikipedia.org/wiki/Maldacena_conjecture (accessed March 10, 2010).

18. The Feynman path integral is a sum over all paths weighted by a mathematical quantity called the "exponential of the action divided by $\hbar$," or

$$\sum_{paths} e^{iS/\hbar}$$

The quantity S is called the *action* (which is a function of the particular path). For the double-slit experiment, there are only two paths to worry about:

(1) electron released from the source and travels through slit 1 and then to point x on the catcher screen (for which we get $F_1 = e^{ikd_1/\hbar}$ where d_1 is the distance from slit 1 to the catcher screen; the action S is simple, just the magnitude of the wave vector k times d_1).

(2) electron released from the source and travels through slit 2 and then to point x on the catcher screen (for which we get $F_2 = e^{ikd_2/\hbar}$ where d_2 is the distance from slit 2 to the detector screen).

So, the amplitude to find the electron on the screen at any point is just $e^{ikd_1/\hbar} + e^{ikd_2/\hbar}$. The probability is just the square of this quantity, $|e^{ikd_1/\hbar} + e^{ikd_2/\hbar}|^2$ (the absolute square, since complex numbers are involved). And if we plot the resulting probability distribution, we get the mysterious interference pattern, in perfect agreement with experiment. It is arising because nature explores all possible paths for the motion of a particle through space and time (in this case, two of them) and adds up the amplitudes for all such paths. The amplitudes interfere when we square to get the probability.

19. There are many textbooks on the subject, such as Charles Kittel, *Introduction to Solid State Physics*, 8th ed. (New York: Wiley, 2004). A particularly simple crystal lattice is a "body-centered cubic lattice." The Wikipedia article can found at http://en.wikipedia.org/wiki/Cubic_crystal_system shows a figure of a body centered cubic lattice (accessed May 25, 2010).

20. In fact, this quantum interference by the scattering of light (x-rays) and other particles off of crystals allows us to detect and measure the crystalline structure.

21. Typically five or more electron volts per electron are required to hop the gap, but this must be delivered over very short distances. It requires very large electrical voltages to cause the electron current to hop into the next band in an insulator (the "breakdown voltage").

22. See, for example, *How Does a Transistor Work?* at http://www.physlink.com/education/askexperts/ae430.cfm and *How Semiconductors Work* at http://www.howstuffworks.com/diode.htm (both accessed May 14, 2010). See also Lillian Hoddeson and Vicki Daitch, *True Genius: The Life and Science of John Bardeen* (Washington, DC: Joseph Henry Press, 2002).

23. David Deutch, *The Fabric of Reality: The Science of Parallel Universes and Its Implications* (New York: Penguin, 1998).

CHAPTER 9

1. We explain the profound connection of symmetry to the laws of physics, including a brief biography of Emmy Noether (see chapter 1, note 1). Also see some of the classics: H. Minkowski, *Space and Time* and A. Einstein, *On the Electrodynamics of Moving Bodies*, both reprinted in the collection *The Principle of Relativity*, edited by Francis A. Davis (New York: Dover, 1952).

2. This is often called the "equivalence of inertial reference frames." It is essentially contained in the *principle of inertia*, as in Newton's first law of motion: "An object will remain at rest or in a state of uniform motion unless acted upon by a force." If an object is at rest, then a uniformly moving observer A (moving relative to B) will see this object to be uniformly moving in the opposite direction; both A and B must conclude the object is not acted upon by a force, hence the equivalence of the physical description in each reference frame of motion. The main point is that both Einstein and Galileo invoke the principle of relativity, but Galileo's symmetry leaves time the same for both observers, while Einstein leaves c the same for both observers.

3. We have provided additional e-appendixes to this book that discuss relativity, gravity, and many other related topics in greater detail in the form of downloadable pdf files, which may be found by navigating our website: http://www.emmynoether.com.

4. Ibid.

5. Some introductory books on general relativity books are Robert M. Wald, *Space, Time, and Gravity: The Theory of the Big Bang and Black Holes*

(Chicago: University of Chicago Press, 1992); Clifford Will, *Was Einstein Right?* (New York: Basic Books, 1993). For the more advanced student: Steven Weinberg, *Gravitation and Cosmology* (New York: John Wiley and Sons, 1972). For the bending of starlight by the sun, observable in total eclipses, see D. Kennefick, "Testing Relativity from the 1919 Eclipse—A Question of Bias," *Physics Today* (March 2009): 37–42; L. I. Schiff, "On Experimental Tests of General Relativity," *American Journal of Physics* 28, no. 4:340–43; C. M. Will, "The Confrontation between General Relativity and Experiment," *Living Reviews in Relativity* 9:39.

6. It's possible to estimate the Schwarzschild radius using Newtonian mechanics. The energy required for a particle of mass m to escape from a massive object is called the "gravitational potential energy," and this is known in the Newtonian theory as $G_N Mm/R$. This is the amount of energy a spaceship, such as *Apollo* 11, with mass m, requires to get far from Earth with radius R, where Earth's mass is M. We now imagine an object of mass m on the surface of a planet, with an enormous mass M, where the gravitational potential energy is so large that it requires that all the mass-energy in the particle be expended to escape. So we simply equate $mc^2 = G_N Mm/R$ and solve for R. We see that m cancels out, and the resulting answer we get is $R = G_N M/c^2$. Now, because we used Newton's theory, we have gotten the wrong answer, but we're only off by a factor of two. The correct answer is $R = 2G_N M/c^2$. Since the mass of the particle trying to escape has dropped out of the formula, any particle of any mass is trapped by such an object, provided its mass M and radius R are related by the above formula. Even light would be trapped at the surface of such a planet.

7. This was written in 1919 and can be found in William Butler Yeats, *Michael Robartes and the Dancer* (Churchtown, Dundrum, Ireland: Chuala Press, 1920). See http://www.potw.org/archive/potw351.html (accessed May 26, 2010).

8. For discussion of gravitational radiation and references see http://www.astro.cornell.edu/academics/courses/astro2201/psr1913.htm (accessed May 17, 2010).

9. Brian Greene, *The Elegant Universe*, Vintage Series (New York: Random House, 2000).

10. Leonard Susskind, *The Cosmic Landscape: String Theory and the Illusion of Intelligent Design* (Back Bay Books, 2006).

11. Ibid.

CHAPTER 10

1. E. J. Squires, *The Mystery of the Quantum World* (Oxford, UK: Taylor & Francis, 1994).

2. Heinz Pagels, *Cosmic Code: Quantum Physics as the Language of Nature* (New York: Bantam, 1984). Each shop has a chief salesperson, skilled in selling his or her version of reality: the boutique where they hawk the latest in string theory, the store selling the many-worlds interpretation, the franchises where visions of quantum computers are on display. Which version of reality shall we buy?

3. N. David Mermin, "Is the Moon There When Nobody Looks? Reality and the Quantum Theory," *Physics Today* (April 1985). The version of Bell's theorem discussed in this section first appeared in this article.

4. Steven Weinberg, *Dreams of a Final Theory: The Search for the Fundamental Laws of Nature* (New York: Pantheon Books, 1992).

5. See *The Stanford Encyclopedia of Quantum Mechanics*, http://plato.stanford.edu/entries/qm-manyworlds/.

6. Paul Davies, *God and the New Physics* (New York: Simon & Schuster, 1984).

7. A. M. Steane, "Quantum Computing," *Reports on Progress in Physics*, no. 61 (1998): 117–73.

8. See Simon Singh, *The Code Book: The Science of Secrecy from Ancient Egypt to Quantum Cryptography* (London: Fourth Estate, 1999).

9. Ibid.

10. Gordon Moore, "Cramming More Components onto Integrated Circuits," *Electronics* 38, no. 8 (April 1865): 4. A pdf version is downloadable at http://download.intel.com/museum/Moores_Law/Articles-Press_Releases/Gordon_Moore_1965_Article.pdf. See also "Martin E. Hellman," http://ee.stanford.edu/~hellman/opinion/moore.html.

11. For an instructive lecture on quantum computing, see "Edward Farhi," http://www.youtube.com/watch?v=gKA1k3VJDq8.

12. Roger Penrose, *The Emperor's New Mind: Concerning Computers, Minds, and the Laws of Physics* (New York: Oxford University Press, 2002); Roger Penrose, *Shadows of the Mind: A Search for the Missing Science of Consciousness* (New York: Oxford University Press, 1996).

13. Francis Crick, *Astonishing Hypothesis: The Scientific Search for the Soul* (New York: Scribner, 1995).

APPENDIX

1. See, for example, Richard P. Feynman, *Lectures on Physics*, vol. 1, chap. 18 (Reading, MA: Addison-Wesley, 2005).

2. We show how this works in the more mathematical discussion in a downloadable e-appendix at our website: emmynoether.com.

INDEX

acceleration, 46
action-at-a-distance, 15, 28, 75–76. *See also* EPR experiment or paradox
Ampere, Andre-Marie, 46
ancient Greeks, 33, 41, 43, 127, 224
Anderson, Carl, 229
Ångström, Anders Jonas; angstrom, physical unit, 72–73
angular momentum, 186, 249, 289–97
antimatter, 228–31, 229 (fig. 33), 240–42, 241(fig. 34)
Aristotle, 41, 43, 286
atom, 16–24, 115–17, 152–78. *See also* hydrogen atom
atomic bomb, 145
atomic mass, 153–78
atomic number 154–78, 155 (fig. 23)
atomic orbitals. *See* orbit

Balmer, Johann, 121
Bardeen, John, 245
Beethoven, Ludwig Van, piano sonata no.15, 89–90
Bell, John; Bell's theorem, 204–17, 273
Bell Labs, 128, 141, 245
Benning, Gerd, 248
blackbody, blackbody radiator, 83–89
blackbody radiation. *See* light
black hole, 255–56
Blake, William, excerpt from "Augeries of Innocence," 270
Bohm, David, 206
Bohr, Niels, 16, 21–22, 24, 29–30, 115–17, 119–47, 188–95, 200–204, 286
Bohr-Einstein debate, 189–204
Boltzmann, Ludwig, 89
Born, Max, 17, 25, 140–47
Bose, Satyendra Nath, 292
Bose-Einstein condensation, 295
bosons, 234–33, 292–97
bound state, 164 (fig. 26), 167 (fig. 27), 175 (fig. 28). *See also* orbit
branes, 266
Brattain, William, 245
Brendel, Alfred, 89–90

c. *See* light, speed of
calculus, 46
caloric, old discarded theory of, 36
Carroll, Lewis, 235–36
Cavendish laboratory, Cambridge, 79, 96, 113, 115
centimeter, definition, 34
CERN, European Center for Nuclear Research, 34–35, 204, 231, 236, 265
chemistry, 151–78
classical determinism. *See* determinism
classical physics, 15, 18, 29, 34, 37, 41, 46, 55–72, 83, 92
cloud chamber, 229–30
color, of light. *See* light, color of
commutation, noncommutation, 131–32
complex numbers, 135–36, 223–25
compound, chemical, 153–63, 173–78
Compton, Arthur Holly, 101–102, 126

Compton scattering, Compton effect, 101–102
concert A, 62–63
conduction band, 242–45
conductors, 244–46
constructive interference. *See* light; waves
Copenhagen (play by Michael Frayn), 144
Copenhagen interpretation of quantum theory, 22, 25, 144–47, 189–95, 203, 275–77, 287
Copernicus, 37
correspondence, principle of, 130
cosmological constant. *See* energy, of vacuum
covalent bond, 175 (fig. 28). *See also* molecule
Crick, Francis, 151, 288
cryptography, 279–81
Curie, Marie, 16
curvature of space-time, 252–54

Dalton, John, 152
Darwinian theory of evolution, 36
de Broglie, Louis, 16, 23, 127–29, 143
de Coulomb, Charles-Augustin, 45
destructive interference. *See* light; waves
determinism, classical; indeterminism, 18, 52, 184, 187, 193
Deutsch, David, 282
de Witt, Bryce, 275
Dickinson, Emily
  "Pattern of the Sun," poem, 43
  "On this Long Storm the Rainbow Rose," poem, 70
diffraction. *See* light; waves
diodes, 245
Dirac, Paul, 16, 142, 225–35, 286
Dirac sea, 227–35, 227 (fig. 32)
DNA, 34, 151
double-slit experiment for light as photons, 102–11, 103 (fig. 15), 105 (fig. 16), 106 (fig. 17), 107 (fig. 18), 109 (fig. 19), 110 (fig. 20)
double-slit experiment for light as waves. *See* Young's double-slit experiment
Douglas, Michael, 268
Duane, William, 102
Duck, Donald, 159

Earth, motion of, 41, 45; natural radioactivity, 159
ECHO satellite, 150
Einstein, Albert, 16–17, 20, 25, 28–31, 37, 48, 94–101, 128–30, 141, 189–204, 249–55, 286
Einstein-Podolosky-Rosen thought experiment. *See* EPR experiment or paradox
electric charge, electric field, 76–81
electrodynamics. *See* electromagnetism
electromagnetism, electromagnetic theory, 36
electron, 14, 21, 24–26, 120–25. *See also* photoelectric effect
electron motion in atom. *See* orbit
electron spin. *See* spin
electron volt. *See* energy
element, chemical elements, 153–78. *See also* atom
ellipse, elliptical orbit, 47, 51
energy
  chemical, 55
  conservation of, 249

definition, 55
discrete levels of orbitals in an atom, 116–17, 121–25, 169–78
$E = hf$, 93
$E = mc^2$, 59, 221–23
electron volt, 121–25
energy bands, in a material, 243
heat, 55
radiation of, 55, 84–94, 121–25
in special relativity. *See* $E = mc^2$
thermal, 84–94
of vacuum, 232–37
entanglement, entangled quantum states, 192–217, 223, 272
EPR experiment or paradox, 28–31, 192–204, 206, 273, 287
ether, in vacuum, 72
Everett, Hugh, 275–77
exchange force, 178–80, 187, 224, 297
exchange symmetry, 178–80, 293–97

Faraday, Michael, 41, 46, 76–78, 152, 286
Fermi, Enrico, 292
Fermilab, Fermi National Accelerator Laboratory, 35, 231
fermion, 292–97
Feynman, Richard, 38, 238–42, 273, 282
Feynman diagrams, 241
Feynman's path integral, 14, 21, 238–42, 273–74
field, 76. *See also* electromagnetism; gravity
Fizeau, Armand, 58
force
of friction, 44
of gravity, 43, 47–53, 251–68
weight, 43
Foucault, Jean, 58
Fourier, Jean Baptiste Joseph, 136–44
Fourier analysis, 136–44, 138 (fig. 22)
Franck-Hertz experiment, 123–25, 124 (fig. 21)
Fraunhofer, Joseph; Fraunhofer lines, 73–76
frequency of light. *See* light; waves
Frost, Robert, "The Lockless Door," poem, 33
Fresnel, Augustin, 71

g, acceleration due to gravity at Earth's surface, 46, 49 (fig. 1), 50 (fig. 2), 51 (fig. 3)
Galilean invariance. *See* relativity
Galileo, Galilei, 14–15, 18–19, 37, 41–46, 57–60, 83, 149, 286
Ganymeade, Jovian Moon, 58
Geiger counter, Geiger tube, 15, 31–32
general relativity. *See* relativity
GeV, giga electron volts. *See* energy, electron volt
Gibbs, Josiah Willard, 89–92
G_N or G_{Newton}, 251
gravitational wave, 257
gravity, gravitational force, 43, 47–53, 232–34, 251–68
Greeks. *See* ancient Greeks
Green, Michael, 264–65
Geene, Brian, *The Elegant Universe*, 265
ground state, 121, 164, 297

h, $\hbar$. See Planck's constant
half-life, 18
Heisenberg, Werner, 16, 18, 21–25, 129–47, 189, 286
Heisenberg's uncertainty principle, 129–33, 142–46

helium atom, 159–62, 168–78, 296–97
Hertz, Heinrich, 16, 41, 80, 96
hidden variable theory, 187–88, 214–15
Higgs mechanism, 32
Hilbert, David, 252
hologram, holography, 237, 266–67
Huygens, Christian, 60
hydrocarbon molecules, 177–78, 178 (fig. 29)
hydrogen atom, 24, 115–17, 121–22, 149–78, 226, 296–97

indeterminacy. *See* determinism
inertia, as frictionless motion, 44
insulators, 243–46
interference fringes, in double-slit experiment, 67
interference of waves. *See* wave
inverse square law force, 47
ionic bond, 175 (fig. 28). *See also* molecule

Josephson, Brian; Josephson junction, 247
Jordan, Pascual, 188

Kepler, Johannes, 37, 149
Kirk, Captain James T. *See Star Trek*
Kohn, Walter, 152

Landscape, 268–70
laser, 17, 295
lepton, 259
LHC or CERN Large Hadron Collider, 34–35, 231, 236, 266
light
  bending or lensing by gravitation. *See* general relativity
  blackbody radiation, 83–94
  as a classical wave, 58–81. *See also* wave
  color of, 21, 55–57, 71, 83–84
  constructive interference, 66–71, 64 (fig. 6), 66 (fig. 7), 67 (fig. 8), 68 (fig. 9), 69 (fig. 10)
  definition of, 55
  destructive interference, 66–71, 64 (fig. 6), 66 (fig. 7), 67 (fig. 8), 68 (fig. 9), 69 (fig. 10)
  diffraction, 66–67. *See also* wave
  $E = hf$, Planck's formula, 93–101
  electromagnetic nature of, 79–94
  emission by atoms, 21, 55, 83–94
  frequency of, 21, 62–71, 87–111, 87 (fig. 13)
  interference, 64 (fig. 6), 66–71, 66 (fig. 7), 67 (fig. 8), 68 (fig. 9), 69 (fig. 10), 108–11. *See also* wave
  as a particle, 58–71, 83–94. *See also* photon
  reflection, 55, 112–13
  refraction, 56, 59, 60 (fig. 4)
  spectrum, 80–81, 80 (fig. 12)
  speed of, 28, 57–58, 71
  thermal emission of, 55–57
  wavelength, 62–81, 87–94, 95–102, 97 (fig. 14)
lithium, lithium fires, 157–58, 168–78, 296–97. *See also* sodium
Lorentz, Hendrik, 250–51
Lorentz transformation. *See* relativity

macroscopic objects, 38
magnet, magnetism
  magnetic field, 76–81, 77 (fig. 11)
  magnetic poles, 76–81, 225
Maldacena, Juan, 237, 266–67

many-worlds interpretation of quantum theory, 275–78
Marconi, Guglielmo, 80
mass, definition of 43, 221–23, 252, 254
Maxwell, James Clerk, 37, 41, 46, 78–81, 89–93, 286
Maxwell's theory of electromagnetism, 78–81, 115
Mendeleyev, Dmitry, 152–63
Mermin, David, 274
meter, definition, 34
MeV, mega electron volts. *See* energy, electron volt
Michelson, Albert A., 250
Michelson-Morley experiment, 250
mode of vibration, 163–65
molecule, molecular bonds, 24, 173–78
momentum, 126, 223, 249, 290
Moore, Gordon; Moore's law, 281
Morley, E. W., 250
muon, 183–89

nanotechnology, 101
negative numbers, 222
neutron star, 297
Neveu, Andre, 262
Newton, Isaac, 14–15, 18–19, 37, 41, 45–53, 59–60, 75, 83, 133, 150, 286
Newton's constant, G_N, 251–52
Newton's laws of motion, 18–19, 46–53, 133–34
Newton's universal law of gravitation, 45–53, 133
node in a vibrational mode, 164
Noether, Emmy, 249
nuclear power, 17
nucleus, of atom, 34, 113–15

Oersted, Hans-Christian, 46
Ohm, Georg Simon, 46
old quantum theory of Bohr, 24, 117
Oppenheimer, J. Robert, 139
orbit
  atomic orbitals, 120–23, 149, 163–78
  collapse of classical orbit in atom, 120
  electron motion in atom, 21, 115–17, 120–23, 149, 163–78
  Keplerian planetary orbits, 120
  molecular orbitals, 173–78
orbital. *See* orbit, electron motion in atom

Pagels, Heinz, 149, 272
parabola, parabolic trajectories, 46–51
particles, 14, 59–71
paths. *See* Feynman's path integral
Pauli, Wolfgang, 17, 119, 129–30, 142, 169–73, 186–87, 286
Pauli poem, by anonymous, from George Gamow's book, 170
Pauli's exclusion principle, 169–73, 187, 287
pendulum, 46
Penrose, Roger, 285
Peoria, Illinois, 28–30
periodic table of the elements, 153–63, 158 (fig. 23), 161 (fig. 24)
phase of a wavefunction. *See* wave
phlogiston, old discarded theory of, 36
photoelectric effect, 96–101, 97 (fig. 14)
photon, 19, 24–27, 83–113. *See also* light
π bond, molecule, 173–76
Planck, Max, 17, 25–26, 89–101, 120, 257

Planck length, or mass, or scale, 234, 257
Planck's constant, 93–94, 116–17, 130, 142–43, 150
Planck's formula, $E = hf$, 93–101, 87 (fig. 13), 116–17, 120–25, 132
Podolosky, Boris, 28, 192
Poe, Edgar Allan, "Pit and the Pendulum," excerpt, 46
Pople, John, 152
positron, 228–35, 229 (fig 33)
powers of ten, 34
probability in quantum theory, 18–19, 21, 25, 31–38, 140–47
Ψ (psi) symbol denoting Schrödinger's wave function, 24–27, 31, 134–43, 184–87
Pythagoras, 127

quanta, quantum particles, 92–115, 184. *See also* photon
quantum computers, 281–85
quantum gravity, 22. *See also* general relativity
quantum mechanics, 70, 203–36
quantum states
  in atoms, 24, 120–25. *See also* orbit; wave function
  collapse of, 22, 29–30. *See also* EPR experiment or paradox
  entangled, 29, 192–217, 223, 272
  of motion, 22. *See also* orbit; wave function
quark, 259
qubit, 280

Rabi, I. I., 183, 204
radioactivity, 18, 183
Ramond, Pierre, 262
reductionist method, 45
reflection, refraction. *See* light
relativity $E = mc^2$, 221
  Galilean relativity, 249–50
  general relativity, 48, 220, 251–57
  Lorentz transformation, 250–51
  principle of constancy of speed of light, 250
  special relativity, 221–27, 249–51
  twin paradox, 250
rogue wave, 136, 138
Rohrer, Heinrich, 248
Römer, Ole, 58
Rosen, Nathan, 28, 192
Rutherford, Ernest, 16, 113–19
Rydberg, Johannes 121
rydberg of hydrogen = 13.6 eV, 121

Schrödinger, Erwin, 16, 20–27, 31–32, 133–47, 151, 286
Schrödinger equation, 18, 20–27, 133–47, 165–69
Schrödinger's cat-in-the-box paradox, 31–32, 277
Schrödinger's interpretation of quantum theory, 25–27, 133–47
Schrödinger's wave function, 23–29, 31, 133–42, 184–87, 243, 293–96
Schwarz, John, 262, 263–65
Schwarzchild radius of a black hole, 255–56
scientific notation, 34
semiconductors, 244–46
Shakespeare, William, excerpt from *Hamlet*, 190
shell. *See* orbit, orbital
Sherk, Joel, 263
Shockley, William, 245
  bond, molecule, 173–76

sodium, sodium fires, 157–58, 168–78. *See also* lithium
solar system, 41, 45
Sommerfeld, Arnold, 129
spectral lines of light, 24, 73–76
spectrometer, 72–76
speed of light. *See* light, speed of
spin, 170–71, 186, 207–14, 224, 289–97. *See also* angular momentum
spinor, 224–26
square root, 222–27
Squires, E. J., 272
Star Trek, 13–14, 17
Steane, Andrew, 278
Stern-Gerlach experiment, 188
Strauss, Richard, 284
string theory, 258–70. *See also* superstring theory
superconductivity, superconductors, 32, 295
superfluid, 32, 295
superstring theory, 36, 263–70
supersymmetry, 249–51, 262
Susskind, Leonard, 268
symmetry, 249–51

Tevatron, Fermilab, 35, 231, 236, 265
theory, definition, 35–36
thermal equilibrium, 84–85
thermal radiation. *See* blackbody radiation
Thompson, J. J., 16, 96, 113–14
transformation, 249–51
transistors, 17, 245
translation, 249
tunneling, 26, 246–48

uncertainty principle. *See* Heisenberg's uncertainty principle
uranium atom, radioactivity of, 18–19, 221–22

vacuum. *See* Dirac sea; ether, in vacuum
vibrational state. *See* quantum states
Victoria's Secret lingerie store, 26, 141, 184, 189, 275

Watson, James, 151
wave
  amplitude of, 61–63, 62 (fig. 5)
  definition of, 23, 61–63, 62 (fig. 5)
  diffraction of, 63–71, 64 (fig. 6), 67 (fig. 8), 68 (fig. 9), 69 (fig. 10), 128
  electromagnetic waves. *See* light
  frequency of, 62–74
  function. *See* Schröedinger's wave function
  gravitational, 257
  interference of, constructive and destructive, 64 (fig. 6), 65, 66–71, 66 (fig. 7), 67 (fig. 8), 68 (fig. 9), 69 (fig. 10), 185, 243
  phase, 65–66, 138 (fig. 22)
  speed of light, 62–63, 62 (fig. 5)
  standing wave, 163–65, 164 (fig. 26)
  traveling wave, 62–63
  wavelength, 34, 61–74, 62 (fig. 5), 163–65
wave function. *See* Schröedinger's wave function
wave mechanics. *See* Schrödinger equation
wave-particle duality, 15
weight. *See* force, of gravity
Weinberg, Steven, 274
Weyl, Hermann, 23
Weyl invariance or Weyl symmetry on strings, 262

Wilde, Oscar, "In the Forest," poem, 118
Wollaston, William, 73
work function, 100. *See also* photoelectric effect

Yeats, William Butler
"The Fish," poem, 231
"The Second Coming," poem, 256
Young, Thomas, 61–62, 66, 102–12, 128, 184–86
Young's double-slit experiment, 66–71, 102–12. *See also* double-slit experiment for light as photons and particles

zero-point motion, 165

Index of Figures

Figure 1: Cannon ball falling, 49
Figure 2: Cannon ball in parabolic trajectory, 50
Figure 3: Cannon ball in orbit, 51
Figure 4: Refraction, 60
Figure 5: Wave-train, or traveling wave, 62
Figures 6a and 6b: Diffraction, 64
Figure 7: Wave interference (or superposition), 66
Figure 8: Interference pattern with double slits, 67
Figure 9: Single slit, absence of interference pattern, 68
Figure 10: Detail of two-slit interference, 69
Figure 11: Dipole field of a magnet, 77
Figure 12: Spectrum of light, 80
Figure 13: Blackbody spectrum, 87
Figure 14: Photoelectric effect, 97
Figure 15: Young's experiment with particle counters, 103
Figure 16: A few photons counted in Young's experiment, 105
Figure 17: Many photons counted in Young's experiment, 106
Figure 18: Photons counted with only one slit open, 107
Figure 19: Photon counting with booby-trap detector, 109
Figure 20: Photon counting with booby-trap detector off, 110
Figure 21: Franck-Hertz experiment, 124
Figure 22: Fourier analysis, 138
Figure 23: Lineup of the atoms (elements), 155
Figure 24: The alkali metals, first column in periodic table, 158
Figure 25: Periodic table of the elements, 161
Figure 26: "Standing" waves on a musical instrument, 164
Figure 27: The atomic orbitals, 167
Figure 28: Molecular orbitals, 175
Figure 29: Some simple hydrocarbon molecules, 178
Figure 30: EPR experimental setup, 197
Figure 31: Bell's experimental setup, 208

Figure 32: Dirac sea, 227
Figure 33: Antiparticle is a "hole" in the Dirac sea, 229
Figure 34: Feynman's view of antimatter, 241
Figure 35: Particle world-line and string world-sheet, 262
Figure 36a: Open and closed strings, 263
Figure 36b: The interaction of two strings, 263
Figure 37: The right-hand rule, 292
Figure 38: Rotations don't commute, 310